ENVIRONMENTAL HEALTH - PHYSICAL, CHEMICAL AND BIOLOGICAL FACTORS

BEACHES

EROSION, MANAGEMENT PRACTICES AND ENVIRONMENTAL IMPLICATIONS

Environmental Health - Physical, Chemical and Biological Factors

Additional books in this series can be found on Nova's website under the Series tab.

Additional e-books in this series can be found on Nova's website under the e-book tab.

ENVIRONMENTAL HEALTH - PHYSICAL, CHEMICAL AND BIOLOGICAL FACTORS

BEACHES

EROSION, MANAGEMENT PRACTICES AND ENVIRONMENTAL IMPLICATIONS

MARGARET CESSA
EDITOR

New York

For permission to use material from this book please contact us:
Telephone 631-231-7269; Fax 631-231-8175
Web Site: http://www.novapublishers.com

LIBRARY OF CONGRESS CATALOGING-IN-PUBLICATION DATA

ISBN: 978-1-63117-239-7

Published by Nova Science Publishers, Inc. † New York

CONTENTS

PREFACE

Sandy beaches are the most abundant coastal systems worldwide, representing approximately ¾ of the worlds' coastlines. In the last decades, these sensitive and dynamic ecosystems have been a target of an increasing human pressure, associated with an unbridled urbanization of coastal areas, caused by high human population growth, but also associated with recreational activities and tourism. Together with global climatic changes and coastal erosion, these processes lead to detrimental impacts and an environmental quality decline of sandy beach ecosystems, imposing an increasing need for adequate assessment and management practices. This book discusses the El Niño-Southern oscillation and coastal response; macrofaunal key species as useful tools in management and assessment practices on sandy beaches; improving coastal knowledge transfer between researchers and managers; the influence of environmental aspects in the variation of natural radioactivity levels in selected Brazilian beach sand samples; nuisance algal blooms; and downcoast erosion triggered by exhaustion of sand supply from sea cliffs.

Chapter 1 – El Niño-Southern Oscillation (ENSO) is a quasi-periodic shift in oceanic and atmospheric circulation that is most evident in the tropical Pacific region, although it affects weather patterns worldwide. Changes in storm tracks, waves, currents, and water level associated with ENSO variability have an important effect on beach erosion and sedimentation along many of the world's coastlines, which should be accounted for in coastal management programs. The warm phase of ENSO (El Niño) is known to cause increased storm wave activity and large losses in sediment volume on beaches along the eastern Pacific, including the West Coasts of North and South America. An increase in mid-latitude cyclone activity and coastal erosion has also been measured during El Niño in the western portion of the Atlantic

Basin, including southern Brazil. El Niño circulation commonly causes a reduction in tropical storm frequency in the North Atlantic, but a possible increase in winter storm activity, which may be important on beaches along the northern Gulf of Mexico. In the western Pacific, particularly Australia, the cold phase of ENSO (La Niña) has been identified as a cause of increased wave energy and resultant beach erosion. Coastal cliffs and coral reefs are also strongly affected. The following chapter will review worldwide research on beach erosion and sedimentation associated with ENSO, and will present new and re-analyzed data from Southern California and the Gulf of Mexico as examples. An understanding of the implications of ENSO is crucial in allowing beach managers to better prepare for episodic sediment loss, particularly within the context of sea level rise and a changing global climate.

Chapter 2 – Sandy beaches are the most abundant coastal systems worldwide, representing approximately ¾ of the worlds' coastlines. In the last decades, these sensitive and dynamic ecosystems have been a target of an increasing human pressure, associated with an unbridled urbanization of coastal areas, caused by the high human population growth, but also associated with recreational activities and tourism. Together with global climatic changes and coastal erosion, these processes lead to detrimental impacts and an environmental quality decline of sandy beach ecosystems, imposing an increasing need for adequate assessment and management practices. In the majority of sandy beaches, the macrofauna is a key element in ecosystem functioning, occupying a central position in the food webs. Some species exhibit abundant populations, representing the bulk of the macrofaunal communities and functioning as key species in these conditions. In this commentary the authors discuss the relevance and utility of macrofaunal key species as tools in beaches integrity assessment, planning of management practices and restoration projects. They start by exposing the reasons behind the selection and use of these organisms as potential tools in assessment and management practices of beaches. Several and alternative methodologies, at the population and at the community ecological levels of organization, are afterwards discussed and their advantages and drawbacks evaluated. Finally, the most relevant key species in worldwide sandy beaches, known in the literature, and with potential as assessment and management tools, are identified. Since several manmade disturbances (e.g., intense human and mechanical trampling, beach grooming with heavy mechanical equipment), and even natural disturbances (e.g., abnormal temperature regimes, frequency and intensity of storms) may cause variations in key species bioecology, with negative impacts perpetuated to the higher ecological levels, monitoring, assessment and management actions focused on this indicator species hold large potential in the fast changing scenario of nowadays, and should be encouraged.

Chapter 3 – It is well recognized that the scientific knowledge on coastal dynamics should contribute to coastal zone management. However, so far, generation of data on coastal processes have been conducted almost exclusively by the research community; the data sets and results are made available in ways that suit that community, but frequently preclude direct application in coastal management. Moreover, there is a plethora of site-specific information that is frequently detained by manager, which is often overlooked by the research community. This chapter assesses both managers and researchers perspective on this subject. Results show that the major gap in knowledge transfer is related with the channels used for data dissemination, which are not adequate to reach managers. This stresses the importance of fostering the existing mechanisms to improve coastal knowledge transfer such as the use of coastal indicators, the existence of linkage agents, the development of "qualitative into quantitative" data tools and the development of managers-oriented tools to support self-data acquisition. Additionally, further meetings and seminars involving researchers and managers - "get together" mechanics - are also suggested by both parties. The perception of coastal researchers and managers concerning gaps in knowledge transfer reflects an unexpected general agreement; this proximity can constitute a major opportunity in bridging these communities together and in the implementation of a two-way route communication approach, thus increasing and improving the role of managers in the coastal knowledge transfer processes.

Chapter 4 – At beaches, environmental implications added to shoreline erosion and artificial interventions on native structure may alter the concentration of the natural radionuclides in certain regions of the coast. In this work, eleven locations at Camburi beach, known to be a naturally high background region, were studied in 2011. The influence of environmental aspects, as geology, average temperatures, variations of the height of the tides and the rainfall levels were related with the variation of ^{226}Ra, ^{232}Th and ^{40}K activities concentrations, during the year 2011 and annual effective dose were evaluated from the ^{226}Ra, ^{232}Th and ^{40}K sands activities concentrations.

Sand samples were monthly collected during 2011, dried, sealed in standard 100 mL HPDE polyethylene flasks and measured by high resolution gamma spectrometry after a 4 week in-growth period. The results, considering samples gamma-rays self-attenuation, show activities concentrations in the range from 4 Bq kg^{-1} to 1380 Bq kg^{-1} for ^{226}Ra, 9 Bq kg^{-1} to 7453 Bq kg^{-1} for ^{232}Th, and 6 Bq kg^{-1} to 504 Bq kg^{-1} for ^{40}K. The absorbed dose rate (D) in outdoor air values laid from 9 nGy.h^{-1} to 5160 nGy.h^{-1} . The annual effective dose values for the sand samples considering a daily exposure of 24 hours all year round laid from 0.05 mSv.y^{-1} to 31.64 mSv.y^{-1}. The annual effective dose values for the sand samples considering daily exposure of one hour all year round laid from 0.002 mSv.y^{-1} to 1.318 mSv.y^{-1}.The annual effective dose (E) for the beach collection points for a resident of Vitoria city, with one hour daily exposure all year round by collection point laid from 0.003 mSv.y^{-1} to 0.232 mSv.y^{-1}. The annual effective dose (*E*) for the collection points from Camburi beach for a tourist with 8 hours daily exposure on January, July or both, laid from 0.004 mSv.y^{-1} to 0.446 mSv.y^{-1}. For a resident of Vitoria city and for a tourist, in the conditions of exposure of this work, all values of *E* are lower than the limit of 1 mSv y^{-1}, recommended for ICRP 60 for the general public. In some months of the year 2011, low average temperatures, low rainfall levels and low variations of the height of the tides can be related with high ^{226}Ra, ^{232}Th and ^{40}K activities concentrations. In April and in October, high rainfall levels and high variations of the height of the tides can be related with low ^{226}Ra, ^{232}Th and ^{40}K activities concentrations.

Chapter 5 – Nuisance algae at beaches are responsible for aesthetic concerns, economic losses and health risks due to production of potent toxins and association with a variety of undesirable bacteria.

Nutrient enrichment with phosphorus, as well as nitrogen in N-limited systems, contributes to eutrophication and development of problem algal blooms. Management of nuisance algae at beaches includes reduction of nutrient inputs from the watershed, reduction of shoreline nutrient sources and shoreline erosion, prompt removal of algal accumulations washed up on shore, education and involvement of the public regarding ways to reduce the problem as well as awareness of health risks associated with algal blooms, and monitoring and timely posting of beach warnings. Use of algaecides is not recommended due to adverse effects on an already stressed ecosystem.

Chapter 6 – The relationship between the protection of sea cliffs from which a large amount of sand was supplied to the downcoast and the resultant downcoast erosion was investigated, taking the northern Kujukuri Beach as an example. A seawall was constructed along with the installation of concrete blocks to protect the downcoast locally, resulting in shoreline recession further downcoast. These actions were recurrently carried out on this coast, and a long stretch of the coastline was covered with concrete blocks, causing the loss of a natural sandy beach. These beach changes were reproduced using the BG model (a three-dimensional model for predicting beach changes based on Bagnold's concept). On the basis of these field data, future measures were considered.

In: Beaches
Editor: Margaret Cessa
ISBN: 978-1-63117-239-7

Chapter 1

El Niño-Southern Oscillation (ENSO) and Coastal Response: An Overview With Examples from Southern California and the Gulf of Mexico

David A. Pepper*

American Public University System, School of Science, Technology, Engineering, and Math, Charles Town, WV, US

Abstract

El Niño-Southern Oscillation (ENSO) is a quasi-periodic shift in oceanic and atmospheric circulation that is most evident in the tropical Pacific region, although it affects weather patterns worldwide. Changes in storm tracks, waves, currents, and water level associated with ENSO variability have an important effect on beach erosion and sedimentation along many of the world's coastlines, which should be accounted for in coastal management programs. The warm phase of ENSO (El Niño) is known to cause increased storm wave activity and large losses in sediment volume on beaches along the eastern Pacific, including the West Coasts of North and South America. An increase in mid-latitude cyclone

* American Public University System, School of Science, Technology, Engineering, and Math, 111 W. Congress Street, Charles Town, WV 25414, David.Pepper@mycampus.apus.edu.

activity and coastal erosion has also been measured during El Niño in the western portion of the Atlantic Basin, including southern Brazil. El Niño circulation commonly causes a reduction in tropical storm frequency in the North Atlantic, but a possible increase in winter storm activity, which may be important on beaches along the northern Gulf of Mexico. In the western Pacific, particularly Australia, the cold phase of ENSO (La Niña) has been identified as a cause of increased wave energy and resultant beach erosion. Coastal cliffs and coral reefs are also strongly affected. The following chapter will review worldwide research on beach erosion and sedimentation associated with ENSO, and will present new and re-analyzed data from Southern California and the Gulf of Mexico as examples. An understanding of the implications of ENSO is crucial in allowing beach managers to better prepare for episodic sediment loss, particularly within the context of sea level rise and a changing global climate.

INTRODUCTION

The term El Niño has been used since at least the 18th Century, when Peruvian fisherman noted an annual warm coastal current in the Equatorial Eastern Pacific. Since it generally peaked near the end of December, they named it in reference to the "Christ Child". However, the term has evolved over time, causing confusion among scientists and the public (Trenberth, 1997). Throughout much of the 20th Century, El Niño was a term typically applied to years when the sea surface temperature increase was considered unusually large. These anomalous years were initially known to be associated with a decline in fisheries, but over time, it became clear that El Niño was also closely connected to atmospheric behavior, including significant changes in precipitation and wind patterns.

Atmospheric circulation over the Pacific has been studied for over a century. According to Walker and Bliss (1932) Swedish meteorologist Hugo Hildebrandsson first identified an inverse correlation between atmospheric pressure in South America and Australia in 1897. Beginning in the early 20th Century, Sir Gilbert Walker statistically analyzed the relationships between pressure and precipitation at different geographical locations, such as the link between flooding on the Nile and the Indian Monsoon. He called the inverse relationship between pressure in the Eastern and Western Pacific the Southern Oscillation (Walker, 1924; Walker and Bliss, 1932). Building upon observations from the 1950's and 1960's, Jacob Bjerknes adopted the term "teleconnection" to describe the link between climatic fluctuations occurring hundreds or thousands of kilometers apart (Bjerknes, 1969). He noted the link between El Niño and the Southern Oscillation, and the joint oceanographic and atmospheric phenomenon thereafter became known collectively as the El Niño-Southern Oscillation (ENSO). In the early 1980's, researchers attempted to identify a sequence of events that could be described as a "canonical El Niño" (Glantz, 2000), although it quickly became clear that El Niño was much more complex than previously thought. Still, El Niño was often treated as a quasi-periodic event, rather than one component of an oscillation. However, in the mid-1980's researchers, notably George Philander, began referring to El Niño's counterpart, La Niña, in which sea surface temperature is particularly low in the eastern Pacific (Philander, 1985). Graham and White (1988) made the case that ENSO is a "coupled oscillator" in the Pacific linked to circulation in the tropical atmosphere, sea surface temperatures in the eastern equatorial Pacific, and ocean surface layer dynamics. Thus, they suggested that El Niño or La Niña were not deviations from a normal "background" state, but the two ends of the same oscillation, which is the dominant view among scientists today.

ENSO escaped widespread public attention until 1982-1983, when the strongest measured occurrence of El Niño up to that time sent global circulation patterns into chaos. The sea surface temperature in parts of the eastern Pacific was more than 3.5°C above normal (University of Illinois, 1999). The South Pacific Ocean had 10 significant tropical cyclones, about twice the annual average (Thompson et al., 1992). Droughts occurred in Australia, flooding devastated parts of South America, and stormy conditions predominated on the West Coast of North America. Another exceptionally strong El Niño phase began in late 1997. Scientists and the public were better prepared for this event, but it had widespread consequences nonetheless.

What began in the 19th Century as a very limited set of observations of an annual coastal current is now considered to be the most important interannual fluctuation in global ocean and atmospheric circulation (Wang et al., 1999; McPhaden et al., 2006). ENSO is known to directly affect the weather on all seven continents, and water level and sea surface temperature throughout the world's oceans. ENSO has been linked to flooding, famine, economic fluctuations, and the collapse of civilizations (Etayo-Cadavid, 2013). El Niño has even been implicated in the worst pandemic in human history, the influenza outbreak of 1918-1919 (Giese et al., 2010). ENSO also has direct and immediate influence on the world's coastlines, which will be discussed in this chapter.

Oceanographic and Atmospheric Characteristics of ENSO

ENSO originates in the Equatorial Pacific region, and follows a very irregular "cycle" with a return interval of 3 to 7 years. ENSO is typically divided into phases: a warm phase, commonly called El Niño, a cool phase, called La Niña, and a neutral phase, in which neither circulation pattern is dominant. Under neutral conditions, the ocean in the eastern Equatorial Pacific is dominated by the cold Peru current and upwelling along the South American coast. Trade winds and Coriolis Force push cold water westward, forming the South Equatorial current, which warms gradually, increasing in temperature by 5-10°C by the time it reaches the western Pacific. The longitudinal gradient in sea surface temperatures generates high atmospheric pressure over the cool eastern Pacific and low pressure in the west, which helps to strengthen the Southeast Trade Winds and reinforce the circulation pattern. During the cold phase, or La Niña, this circulation pattern is stronger than normal, resulting in a larger temperature and pressure gradient between the eastern and western Pacific, along with stronger Trade Winds and a magnified South Equatorial Current. During a typical warm phase, or El Niño, this pattern breaks down. Warm water pools in the eastern Pacific, generating lower atmospheric pressure than normal. The Trade Winds and the South Equatorial Current weaken, and the Equatorial Countercurrent, located a few degrees of latitude to the North, may strengthen, bringing warmer water eastward (Wyrtki, 1973). Although the pressure gradient in the Pacific does not completely reverse, the low pressure region typically found in the west typically shifts toward the central Pacific, and a secondary low pressure region sometimes forms off the South American coast.

These changes in wind, pressure, sea surface temperature, and surface currents are accompanied by changes in local sea level and precipitation. During non-ENSO conditions, sea level is about 0.5 meters higher in the western Pacific than it is in the east, whereas the difference between the two sides of the Pacific may be negligible during the El Niño phase (NOAA, 2013a). During neutral and La Niña intervals, the central Pacific and the western coast of South America receive low to moderate precipitation, whereas heavy precipitation often accompanies the El Niño phase in these regions. Conversely, El Niño may cause drought in areas normally affected by the summer monsoon, such as Southeast Asia and Northern Australia, whereas high rainfall during La Niña can cause flooding in these regions.

The specific trigger for the shifts associated with ENSO is not well understood, although in a very general sense it appears to be a heat-balance mechanism in the Equatorial Pacific Ocean. Palumbo (2001) argued that intense El Niño events can be predicted with some accuracy since the Pacific warm pool must release heat energy at a roughly quantifiable long-term rate; therefore if heat is not released in small amounts during frequent, less significant El Niño episodes, it will eventually be released during a more intense one. Fairbridge (1998) offered another perspective. He argued that the origin of "super" El Niño events may lie beyond the Earth itself, coinciding with long-term cycles of 69, 104, and 208 years, and short-term cycles of 2 to 9 years in planetary and lunar orbits, along with solar flares. He linked these return cycles with beach-ridge sequences in the Arctic, including Canada, Scandinavia, and Russia, and lower latitude environments, including the coasts of California, Northern Australia, and the Gulf of Mexico. Nonetheless, the oceanic and atmospheric components are so strongly interlinked that they are virtually impossible to separate--wind drives surface currents, surface currents influence sea surface temperature, sea surface temperature influences pressure, and pressure generates wind. It's a classic positive feedback loop. These interactions generate other atmospheric and hydrodynamic responses that ultimately affect the coast (Figure 1). Although our predictive capabilities are increasing, there's still no surefire way to predict when a particular ENSO phase will begin, how strong it will be, or when it will end. At the moment, fairly reliable forecasts are limited to a few months in advance (NOAA, 2013b).

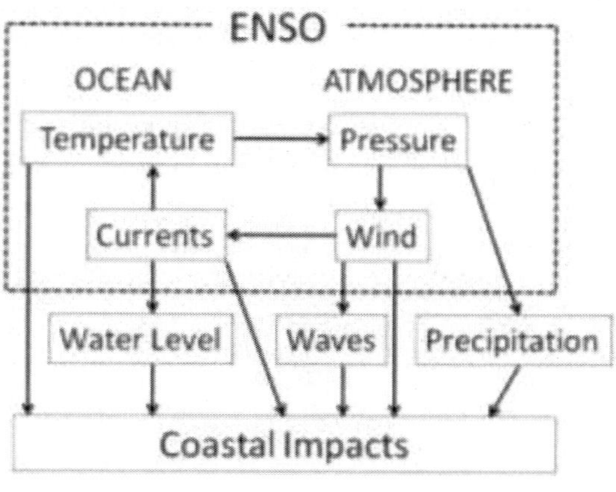

Figure 1. Oceanic, atmospheric, hydrodynamic and coastal relationships associated with ENSO.

QUANTIFYING ENSO

ENSO is by definition an oscillation between two "extreme" phases known as El Niño and La Niña, with an infinitely large set of possible atmospheric and oceanic states lying in between. El Niño and La Niña phases themselves vary in intensity, and there is no inherent upper limit to how strong these occurrences can be. It is important, therefore, to define ENSO quantitatively so that we can accurately identify phases of atmospheric circulation and compare their magnitudes through time. Scientists express the ENSO state most simply by using the Oceanic Niño Index (ONI) and the Southern Oscillation Index (SOI). Both use the concept of departures from normal measurements at a variety of stations, typically calculated using a three-month moving average. The ONI compares three-month averaged temperatures measured in an area of the central Pacific known as the "Niño 3.4 Region" with a baseline average temperature measured in the same region between 1971 and 2000 (Dahlman, 2009). Temperatures that are 0.5°C above this 30-year average indicate El Niño circulation, while temperatures 0.5°C below average indicate La Niña conditions. The Southern Oscillation Index is calculated using the mean monthly normalized sea level pressure difference between Tahiti (central Pacific) and Darwin (western Pacific) (Chen, 1982). Negative values indicate El Niño circulation and positive values indicate La Niña. A more complex but arguably more informative measurement is the Multivariate ENSO Index (MEI) introduced by Wolter and Timlin (1993). This measurement relies on six key variables: atmospheric pressure; zonal and meridional winds; sea surface temperature; air temperature; and cloudiness (Earth Systems Research Laboratory, 2013a). Positive values indicate El Niño and negatives indicate La Niña (Figure 2). The duration of El Niño or La Niña conditions is also a consideration in classifying the phase of circulation. Trenberth (1997) suggested that a five-month running mean in sea surface temperature should be used to identify significant El Niño or La Niña events.

On the basis of these MEI, numerous significant phases of El Niño have been identified since 1950, including 1957-58, 1965-66, 1972-73, 1982-83, 1991-92, 1997-98, 2009-10, of which 1982-83 and 1997-98 were particularly strong (Wolter and Timlin, 1998; Earth Systems Research Laboratory, 2013a). La Niña phases include 1949-51, 1954-56, 1964-65, 1970-71, 1973-75, 1988-89, and 2010-11, of which 1954-56 and 1973-75 were the strongest (Earth Systems Research Laboratory, 2013a).

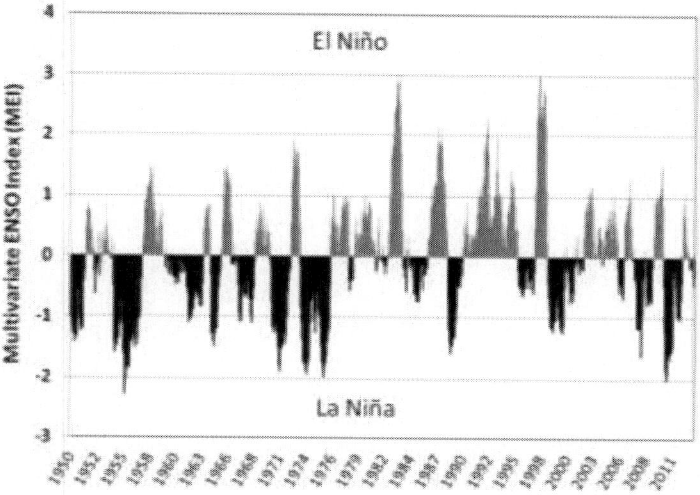

Figure 2. The multivariate ENSO index (MEI), 1950-2013.

ENSO AND ITS VARIATIONS

As scientists have developed a more sophisticated understanding of ENSO, the simple linear nature of the oscillation between two extremes has been questioned. Some researchers have suggested that another "type" or "flavor" of El Niño can occur in which warming is concentrated in the Central Pacific rather than in the East. This circulation pattern has been called Warm Pool El Niño (Kug et al., 2009), the Central Pacific (CP) El Niño, or El Niño Modoki, from a Japanese word meaning "a similar but different thing" (Ashok et al., 2007). Although this pattern has been shown to have similar origins in subsurface circulation (Ramesh and Murtugudde, 2013) it manifests itself somewhat differently from the canonical, Eastern Pacific (EP) El Niño type. According to Weng et al. (2007) El Niño Modoki may be associated with an increased number of typhoons in the Western Pacific and hurricanes in the North Atlantic, as well as drought in western North America. These impacts are essentially the opposite of what might be expected under typical El Niño conditions. On the other hand, researchers have measured coastal impacts in North America during El Niño Modoki that are comparable to those experienced during a typical El Niño (Barnard et al., 2012), so it is unclear how El Niño Modoki might be incorporated into a generalized classification of ENSO.

ENSO is also closely tied to other oscillations in ocean and atmospheric circulation over longer time scales, most notably, the Pacific Decadal Oscillation (PDO), which is also sometimes called the Interdecadal Pacific Oscillation (IPO). As its name suggests, the PDO occurs over periods much longer than ENSO, typically over the course of 10-30 years. Some of the changes occurring during the PDO are similar to those measured during ENSO. During the warm phase, sea surface temperature increases in the eastern Pacific, including considerable warming along the North American Coast. Temperature decreases in the eastern Pacific during the cool phase. There is strong evidence that El Niño/La Niña episodes may be intensified when they are in phase with a warm PDO/cold PDO (Chan, 2005; Yoon and Yeh, 2010), but more work needs to be done to better determine the details of this relationship.

ENSO and Climate Change

There is ample evidence that ENSO has been an important part of Earth's climate system for thousands, and probably millions, of years. Rosenthal and Broccoli (2004) state that ENSO circulation has been active throughout the Holocene and may have been influential during the Last Glacial Maximum. McPhaden et al. (2006) noted that a warm period occurring 3-5 million years ago may have been associated with permanent El Niño conditions, and that the Earth experienced increased temperatures and a more energetic ENSO cycle 35-55 million years ago.

The link between ENSO and climate change is an important issue, although one that has not yet been answered definitively. El Niño has been described as heat balance mechanism in the Equatorial Pacific, so it would seem logical that global warming would cause ENSO cycles to intensify. Proving and quantifying this relationship has been tricky, however. Fedorov and Philander (2000) asked the question "Is El Niño Changing?" and concluded that it is possible that due to global warming, the answer may be "yes". Cobb et al. (2013) demonstrated that ENSO variance has been higher than average during the 20th Century, based on 7,700 years worth of oxygen isotope data from coral reefs. They commented, however, that it is difficult to determine whether the cause has been global warming, either anthropogenic or natural. McPhaden et al. (2006) were quick to point out that the shift toward El Niño conditions since 1970 may not be linked to warming, despite the relationship between warming and ENSO in the past. They also noted that it remains uncertain how climate change will affect ENSO in the future. According to Stevenson et al. (2012), computer models show that the response time of the coupled ocean-atmosphere system in the tropics may be too slow to show significant change as a result of increased CO_2 concentrations during the 21st Century. They did predict, however, that many of the regional effects of ENSO could intensify, especially during the boreal winter. At the moment, it appears that there is simply insufficient evidence to make conclusive predictions of ENSO's future behavior.

GLOBAL ENSO EFFECTS

Although ENSO is focused in the Equatorial Pacific, it is arguably the most significant cause of global variability in oceanic and atmospheric circulation on an inter-annual to decadal time scale. The effects of El Niño are most noticeable on a global basis during the boreal winter. During El Niño years, winter weather tends to be warmer in Japan and Korea, along with large sections of Alaska, and parts of both Western and Eastern Canada and the northern United States. Wetter areas include Equatorial regions in the Central Pacific, East Africa, and Western South America, as well as coastal mid-latitude regions in North and South America. Drier than average regions include large portions of the Western Pacific, Southeast Asia, and Australia, and eastern portions of Southern Africa and Equatorial South America. During the Northern Hemisphere summer, many of these dry areas persist, and the Indian monsoon tends to weaken, leaving South Asia drier than usual. La Niña years produce conditions that are basically the opposite of those experienced during El Niño. (National Weather Service, 2012)

The influence of ENSO on weather changes in North America have been well-studied and are very influential on the dynamics of the Pacific coast. Although the direct effects of the Southern Oscillation are felt only within a few degrees of the Intertropical Convergence Zone, the associated outflow of mass and energy initiates a wave-like disruption that extends toward the poles (Earth Systems Research Laboratory, 2013b). During neutral or La Niña winters, high pressure predominates over the North Pacific, with low pressure forming over Northern Canada. This configuration tends to steer mid-latitude cyclones toward British Columbia and the Pacific Northwest. During El Niño winters, however, persistent low pressure in the Pacific shifts or sometimes splits the jet stream and steers storms toward the California Coast (Storlazzi and Griggs, 2000). This shift in the paths of winter storms is ultimately one of the most significant influences of ENSO on coastal dynamics.

ENSO and Coastal Impacts

ENSO affects the world's coastlines in a variety of ways, some of which are quite subtle, while others are very dramatic. Moreover, some effects are more predictable than others. The most important changes fall into three general categories: 1) Hydrodynamics and wind 2) Precipitation and inland runoff 3) Sea surface temperature. The first two of these are primarily affected by changes in the frequency, path, and characteristics of storms that influence a specific geographical area, although changes in the Trade Winds do influence large-scale currents and water level in the tropics. Changes in sea surface temperature have the most notable effect on coastal biology, particularly coral. The primary focus of this chapter is categories 1 and 2, although a brief note regarding corals is included.

Tropical Cyclones

Tropical cyclones, including hurricanes and typhoons, have an enormous impact on coastlines when they make landfall. While a comprehensive discussion of the impacts of tropical cyclones is beyond the scope of this chapter, they are commonly associated with flooding, sediment redistribution, barrier island migration, wetland loss, shoreline retreat, and damage to coastal infrastructure. Thus, any effects of ENSO on tropical cyclone frequency are significant to long-term coastal development. However, tropical cyclone landfalls on any specific stretch of coastline are fairly rare, even in comparatively active years, so the ENSO circulation pattern at any particular time should be treated in probabilistic rather than deterministic terms.

Tropical cyclone formation depends upon three key factors: 1) Sufficient Coriolis force to initiate storm rotation 2) Sea surface temperature in excess of 26.5°C 3) Low wind shear in the upper atmosphere. Obviously, the first of these, Coriolis force, is not affected by ENSO; its primary influence is to inhibit tropical cyclone formation within a few degrees of the Equator. On the other hand, wind shear and sea surface temperature in the Pacific, Indian, and Atlantic Oceans are affected by ENSO.

Chu (2004) summarized the influence of ENSO on tropical cyclone occurrence in the Pacific and Atlantic Oceans. In the North Atlantic, El Niño episodes are associated with fewer and somewhat weaker storms, which often form further north than in neutral years. ENSO has a very significant effect on landfall rates of major hurricanes in the United States. Bove et al. (1998) found that the probability of a major hurricane making landfall during an El Niño year was only 23%, compared to 58% during neutral circulation, and 63% during La Niña. In fact, two or more major hurricanes made landfall in 27% of the La Niña years included in their 98-year study period, but never during an El Niño year.

The North Pacific, unlike the Atlantic, has two distinct areas of tropical cyclogenesis. The eastern zone lies off the coast of Mexico and produces hurricanes that most commonly travel westward and dissipate in the central Pacific without making landfall. The western zone produces more cyclones than anywhere else on Earth; these typically strike Southeast Asia, China, and Japan, where they are known as typhoons. During El Niño years, both areas of tropical cyclogenesis shift toward the central Pacific, and storms in both zones increase in duration. Storms in the east become more intense, while changes in storm track are notable in the western basin. There is also a reduction in tropical cyclone frequency in the western North Pacific in the year immediately following a warm ENSO event. Zhang et al. (2012) found that tropical cyclone patterns are slightly different in East Asia depending upon the type of El Niño. Of particular note, during Central Pacific El Niño conditions, tropical cyclones are more likely to make landfall in Japan and Korea, particularly during the summer.

Tropical cyclones in the South Pacific generally form farther east during El Niño years, causing storm frequency to increase in the southeastern Pacific and decrease in the southwestern Pacific. As a result, cyclone landfalls are more common in Australia during La Niña years, whereas areas such as French Polynesia and even Hawaii experience more during El Niño (Chu, 2004). Considerably less is known about the impact of ENSO on cyclone formation over the Indian Ocean, although Sumesh and Ramesh Kumar (2013) suggest that there is likely a significant connection, particularly during El Niño Modoki.

Thus it appears that during El Niño episodes, the central Pacific and parts of East Asia, notably Japan and Korea, may be more susceptible to tropical cyclone landfalls. Conversely, the western Atlantic, including eastern North America and the Caribbean, as well as parts of the South Pacific, including the Australian coast, may be more susceptible during La Niña. However, because of the probabilistic nature of tropical cyclone landfalls, it is difficult to be very specific regarding the implications for coastal response or coastal zone management.

Winter Storms

The impacts of ENSO are considered to be most significant during boreal winter, when El Niño circulation is typically the strongest. This is also the time of year when coastlines in the mid-latitudes are affected by various types of winter storms, including mid-latitude cyclones and frontal passages. Unlike tropical cyclones, which are much less frequent, winter storms impact most coastlines numerous, sometimes dozens, of times per year. Mid-latitude storms occasionally approach the intensity of tropical cyclones, but are normally much less violent, sometimes involving little more than light rain and a shift in wind direction. These storms are quite variable in size and can in some instances extend more than a thousand kilometers. Winter storms are a key agent of change on many coastlines, such as North and South America, and Australia, because of their high frequency and large size. ENSO strongly influences the track of winter storms, and thus, the frequency and intensity of storms experienced along many coastlines. Perhaps nowhere is the effect of ENSO on winter storms more evident than on the West Coast of North America.

North America: West Coast

As discussed in an earlier section, the strength and position of quasi-stationary high and low pressure systems over the North Pacific and Western North America change appreciably during El Niño phases of circulation, typically steering winter storms further south than in non-El Niño years. The coasts of northern Mexico, California, and Oregon are impacted by more frequent and more powerful storms as a result. These storms have been shown to cause significant change on beaches and cliffed coasts, along with damage to coastal infrastructure.

Komar (1998) noted that in the Pacific Northwest, El Niño increases water level, storm-generated wave activity, rip-current strength, and longshore sediment transport, all of which cause increased erosion. Erosion may be enhanced by migration of rivers and tidal inlets as well as the effects of coastal structures such as jetties and breakwaters. He reported an increase in water level of about 0.5 meters above astronomical tides, and maximum storm waves with a deep-water height of nearly 10 meters in January, 1998, (the return interval of waves of this magnitude is 10-20 years in the area). Further work documented the role of the Netarts Littoral Cell in this pattern of erosion and deposition (Revell et al., 2002). Allan et al (2003) stressed the need for coastal managers to understand that erosion is episodic on the Oregon coastline. They noted that El Niño intervals are particularly important in causing short-term erosion, especially at the southern end of littoral cells, north of obstructions such as cliffs and jetties. Later, Allan and Komar (2006) compiled observations from multiple sites along the West Coast of the United States to extend their conclusion that El Niño years are associated with increased water level, storm wave activity, and coastal erosion. They noted that when measuring on this extended scale, there was some variation in beach response with latitude and differing local characteristics.

Using LIDAR, Quan et al. (2013) found that large-scale erosion rates increased significantly in Monterrey Bay during the El Niño events of 1997-1998 and 2009-2010 as compared to non-El Niño intervals. They also demonstrated that, on a smaller scale, hot spots shifted at regular intervals of 160 meters between El Niño and non-El Niño years. Revell et al. (2011) showed that the effects of strong El Niño episodes may persist for several years on coasts dominated by unidirectional longshore transport. They documented the formation of four erosion hot spots on California beaches during the 1997-98 El Niño episode; beaches updrift of these hotspots continued to accumulate sediment for more than three years, while downdrift locations continued to erode for more than six years.

Storlazzi and Griggs (1998, 2000) statistically analyzed 86 years of meteorological and oceanographic data to assess the effects of ENSO along the Central California coastline. They showed that 75% of the damaging storms along the Central California Coast occurred during El Niño episodes, with an average of 1.2 damaging storms during El Niño years and only 0.3 during non-event years (Storlazzi and Griggs, 1998). They found that there was a nonlinear correlation between ENSO intensity and wave height, sea level, and precipitation in Central California, thus demonstrating that particularly intense ENSO episodes are the most significant drivers of long-term sea-cliff erosion (Storlazzi and Griggs, 2000). They argued that ENSO is one of the primary factors in overall coastal evolution in the region (Storlazzi and Griggs, 1998).

Cliff retreat during El Niño episodes is a particularly significant occurrence because it is irreversible, unlike sediment loss from beaches, which can be regained during fair weather. The USGS conducted extensive monitoring in Montara State Beach and Pacifica, California during the El Niño winter of 1997-1998. Retreat of up to 14 meters (more than fifty times the average annual rate) was measured on cliffs at the southern end of pocket beaches. This was accompanied by sediment loss on beaches in front of retreating cliffs. Cliffs at the northern end of beaches tended to be stable, and accumulated significant sediment deposits (USGS, 2013a; 2013b).

Of course, all El Niño episodes are not created equal. Storlazzi et al. (2000) conducted a comparison of the impacts of the El Niño episodes of 1982-83 and 1997-98 on the Central California coastline. Although the later episode was more intense by most measures, they found that accompanying economic loss was only about half of that incurred in 1982-83. The authors attribute this to both natural and human factors. In 1982-83, the most energetic waves occurred during periods of increased set-up and offshore flow at the coast, caused by high tides and strong southerly winds. In 1997-98, high waves coincided with low tides and northwesterly winds, causing onshore flow. Protective structures were also more substantial in 1997-98, at least partially in response to the damages that had occurred during previous El Niño episodes. This is a promising finding, indicating that coastal managers and residents were cognizant of the problem, even a decade ago, and were able to adjust with some effectiveness.

It appears, also, that variations in El Niño "type" may not cause noticeably different responses on the West Coast of the United States. Barnard et al. (2011) found that increases in wave energy and water level, along with rates of shoreline retreat during the 2009-2010 El Niño Modoki were similar in California to the 1997-1998 El Niño. The rate of shoreline retreat in Oregon and Washington, however, was higher than in 1997-1998, particularly just north of barriers to longshore transport, as a result of an increase in southerly wave energy.

Trends in ENSO and coastal erosion in recent decades are a cause for concern. Adams et al (2008) used wave hindcasting to demonstrate that the influence of El Niño storm waves in Southern California has increased since the 1960's. Some of this increase may be attributable to the predominance of a warm phase of the Pacific Decadal Oscillation, which they found combined with El Niño to produce waves that were higher, had longer periods, and were more westerly than during cold phases of the PDO. Hapke et al. (2009) conducted a comprehensive survey of changes along California coastlines over periods of up to 120 years. The researchers noted that 66% of sandy coastlines experienced net erosion during the most recent short-term interval of data availability (~25yrs.), as opposed to a long-term figure of only 40%. They suggest that this attributable to the occurrence of intense El Niño episodes during the 1980's and 1990's.

Although El Niño has clearly been shown to cause erosion over annual, and even decadal time scales, the long-term effect of El Niño on beaches along the West Coast is far from clear. El Niño winters are three times more likely to produce storms that cause coastal erosion in central California as other years, but many beaches have been observed to recover quickly. Gary Griggs commented in ScienceDaily (2000) that El Niño years are important in contributing to the coastal sediment supply in California since increased storm activity produces greater stream runoff.

The importance of El Niño to long-term sediment supply has been corroborated by others. Inman and Jenkins (1999) showed that sediment flux in small California rivers can be 27-times higher during severe El Niño years than during dry years, noting that the climate in California shifted abruptly in 1969 from a very dry regime to a much wetter one, with a considerably greater sediment supply. Hudson (2003) cited the significance of ENSO in episodically increasing stream discharge and suspended sediment transport in Eastern Mexico. He noted that although El Niño episodes are typically thought to be more significant along much of the Gulf of Mexico coast, La Niña generates more sediment discharge in some areas. Andrews et al. (2004) noted that the frequency of flooding for a given exceedance probability in California coastal streams increased tenfold in the southern part of the state during El Niño years, although it decreased somewhat in Northern California. Heavy precipitation, it should be noted, may also enhance erosion in some environments. According to Storlazzi and Griggs (2000) high stream discharge may inundate shallow lagoons and estuaries, while coastal bluffs may become more vulnerable to collapse as saturation increases pore pressure, decreases inter-particle friction, and increases the weight of their constituent materials.

Masters (2006) provided an interesting long-term perspective on California coastal evolution. She used fossil proxies to reconstruct episodes of erosion and accretion on the Southern California coast during the Holocene. She found that changes in sediment abundance and wave energy associated with ENSO have been important influences on the coastal geomorphology of the region during the past several thousand years. She suggested that active El Niño periods were associated with erosion of exposed beaches due to increased wave energy, but also with accretion on more sheltered beaches, where river runoff contributed to an increased sediment supply. Thus, it appears that on a long-term basis, ENSO plays an important role in both the erosion and accretion of the California Coast. Over time scales longer than centuries, mechanisms such as subsidence, tectonism and eustatic sea level change may dominate, while the effect of ENSO is "a wash".

A Brief Case Study from Southern California: Hermosa Beach

The city of Hermosa Beach is a coastal community with approximately 20,000 residents, located 27 km southwest of downtown Los Angeles. It is a popular destination for recreation and tourism, and the community comprises a real estate market worth hundreds of millions of dollars. The beach itself is roughly three kilometers long and 200 meters wide, and consists predominantly of medium to coarse quartz sand with a mean diameter of approximately 0.6 mm. The beach is delimited clearly on the landward side by a one-meter high concrete wall that separates it from a concrete walkway, known as the Strand. Hermosa Beach occupies the southernmost three kilometers of a continuous eight-kilometer stretch of sandy beach, terminating in the south at the Redondo Beach/King Harbor Marina Complex. The beach includes a 378 meter-long fishing pier mounted on concrete pilings that is located roughly one kilometer from the beach's south end, as well as six storm drains with concrete outlets that are visible when the sand level is low. There is a permanent lifeguard station adjacent to the pier and twelve to fourteen movable lifeguard stations located at approximate 200-meter intervals.

The geology of the region is quite complex, having been influenced by folding, faulting, volcanism, and submergence throughout its history. The Los Angeles Basin and its bordering mountains incorporate igneous, metamorphic, and sedimentary rocks ranging in age from over one billion years to less than one million years. The majority of the coastline consists of sandy beaches, which are variously backed by bluffs, dunes, or urban development. The predominant natural sediment sources for the bay are river runoff from the Santa Monica Mountains and onshore transport from the inner continental shelf. Nourishment projects that have occurred since the mid-1900's, however, are the largest sand source for most of the coastline (Griggs et al., 2005). Longshore transport in the region is dominated by the Santa Monica littoral cell, which moves sand southward from the beaches and feeder streams in the north and along the coast to the southern tip of the bay, where it flows into Redondo Canyon, and is lost to the continental shelf. Longshore transport is interrupted somewhat at various locations along the coast, particularly at Marina del Ray and King Harbor, where breakwaters and jetties have been constructed to preserve shipping and boating activities. Smaller structures, most notably groins, have also been built at numerous locations along the coast.

Hermosa Beach lies toward the south end of Santa Monica Bay, a crescent-shaped embayment in the Pacific Ocean that stretches 32 kilometers between the Santa Monica Mountains to the North, and the Palos Verdes Hills to the South. Headlands largely protect the beach from northwesterly swell. Hydro-meteorological influences include occasional Santa Ana Winds during the fall, which decrease water level and suppress onshore wave activity for a few days; distant tropical storms, which generate southwesterly swell; and wintertime mid-latitude cyclones and frontal systems, which generate strong winds, rainfall, and short-period storm waves. Mid-latitude storms are most common between November and March, while fair weather and low-frequency swell dominate during the remainder of the year. As a result, the beach profile follows a typical summer-winter cycle, with discontinuous nearshore bars generally forming 20-100 meters from shore (measured from mean water level) during the winter.

Figure 3. Hermosa Beach Pier, lifeguard station, and protective berm with beach scarp in foreground.

The beach is heavily managed. Bulldozers and tractors are used regularly to sift, level, and redistribute sand. The California Department of Beaches and Harbors prepares for winter storm season on an annual basis by bulldozing sand from the backshore area to create a sandy berm approximately four meters high and two-hundred meters long, to protect the lifeguard station and other infrastructure near the pier (Figure 3). It also responds to individual storms by bulldozing sand from the front of storm-drainage pipes which occur at intervals every few hundred meters along the beach, creating artificial channels from the pipe outlet to the ocean (Figure 4).

Figure 4. Bulldozed drainage channel following a winter storm.

Detailed observations and photographs, along with a somewhat limited set of GPS measurements, were collected during more than thirty visits to a one-kilometer stretch of Hermosa Beach during the strong El Niño winter of 2009-2010. Large changes in beach morphology were observed during this time, particularly during a series of storms that occurred during the last two weeks of January. Large beach scarps formed and beach volume decreased as a considerable quantity of sand was transferred offshore. This exposed large deposits of gravel and cobble-sized igneous sediments on the foreshore. These sediment deposits, despite their large size, were surprisingly ephemeral, appearing and disappearing within days (Figures 5a,b). The occurrence of these deposits was not uniform or simultaneous along the beach; in fact deposits apparently shifted southward along the shoreline by tens or hundreds of meters over the course of days. These shifts may have been the result of alongshore movement of gravel and cobbles, but is more likely related to the episodic erosion, alongshore transport, and deposition of sand, alternately exposing and uncovering the deposits. Surprisingly, the appearance of rocky deposits seemed to only weakly linked to individual storms; however, the development of a winter-storm beach profile over the course of the storm season appeared to be a fundamental factor in exposing vast quantities of gravel and cobbles. By the summer of 2010, the beach had fully recovered to its typical summer profile. During the subsequent winters of 2010, 2011, and 2012, sediment lost from the beach profile did expose rocky deposits, but the magnitude of exposure did not approach that seen during the El Niño episode.

(a)

(b)

Figures 5a,b. Deposit of gravel and cobbles present on February 19, and absent on February 22.

Figure 6. Artificial berm with Hermosa Beach lifeguard station following a winter storm.

Qualitative observations from Hermosa Beach suggest that beach change associated with El Niño was temporary. In fact, in some areas where rocky deposits were exposed, partial "recovery" apparently took only days, and adjustment to a normal summer profile was complete within weeks or months. The author is aware of no reports of infrastructure loss due to wave activity or water level rise, and the most exposed building, the lifeguard station, was well-protected by the artificial berm (Fig. 6). Beach usage is always much lower during the winter than it is during the summer, and there is no evidence to suggest that the loss of sand and exposure of coarse material lowered it further. In fact, the presence of rocky deposits attracted many beachgoers, who were commonly observed combing through them to find attractive rocks to use as home and garden decorations. Along beaches, such as Hermosa, that benefit from appropriate management strategies and adequate sediment supply, ENSO may ultimately be little more than a short-term change in scenery.

South America

In South America, response to ENSO has been documented on both Pacific (west) and Atlantic (east) coasts. Warm water along the Equatorial Pacific coast is the oceanic signature of El Niño, and it is characteristically accompanied by reduced barometric pressure in the region. Low pressure is associated with enhanced thunderstorm activity and increased precipitation along South America's West Coast, which has been linked to enhanced sediment discharge from inland drainage. For example, Sandweiss et al. (1998) documented increased transport to the Peruvian coast, leading to the formation of gravel beach-ridge sequences at several sites.

On the other hand, increased storminess is associated with coastal retreat. Federici and Rodolfi (2001) studied stretches of the Ecuadorian Coast that were experiencing long-term erosion ultimately caused by sea level rise and human activity. Cliff retreat accelerated greatly during El Niño episodes, becoming so severe that the authors declared that coastal protection strategies were unrealistic. According to Martinez et al. (1994) "Understanding El Niño is recognized as a key to safe development of the coasts of Peru, Ecuador and Colombia."

Along the eastern coast of South America, the effect of El Niño is more variable, although increased storminess is common in Southern Brazil and Northern Argentina. Lamour and Soares (2003) noted that little net change in shoreline position occurred on four Brazilian beaches during the successive La Niña and El Niño events, despite changes in winter storm characteristics. However, de Souza and Angulo (2003) stated that El Niño is associated with intensification of frontal systems in Southern Brazil, which increases south and southeasterly storm wave activity, causing strong cross-shore transport and significant beach retreat. Faraco et al. (2006) found that El Niño causes fronts to remain stationary for long periods over the Brazilian coast, allowing easterly storm waves to develop and mobilize sediment. They further noted that coastal orientation is a key factor in beach response, since dominant wave direction is different during El Niño occurrences than it is during La Niña. Esteves et al. (2006) observed that ENSO is associated with shifts in longshore sediment transport direction along the Brazilian coast. Longshore currents most commonly flow toward the southeast in neutral years, but tended to flow toward the north or northwest during El Niño and La Niña years. Areas of erosion and accretion occurred during all circulation types, but the change in longshore transport direction shifted the locations of these areas between El Niño, La Niña, and neutral episodes. The authors suggested that interannual transport may ultimately be balanced over longer, possibly decadal, time scales.

ENSO also has an effect on aeolian transport as well. Maia et al (2005) found that in Northern Brazil, El Niño causes a northward displacement of the

Intertropical Convergence Zone and a more pronounced dry season, leading to an increase in dune migration.

Australia and New Zealand

In Eastern Australia, and to some extent, New Zealand, coastal responses to ENSO are very significant and quite well-documented. The effects of ENSO are basically the reverse of what occurs on the West Coast of North America. Erosion tends to occur during La Niña years, when there is a higher storm frequency, whereas accretion is often associated with El Niño (Brooke, 2013). Research in New South Wales (Phinn and Hastings, 1992) and Queensland (Phinn and Hastings, 1995) has shown that mean deepwater wave energy is significantly lower during El Niño episodes than during neutral or La Niña phases, possibly owing to a decrease in southwestern Pacific tropical cyclone activity. You and Lord (2008) found a direct correlation between yearly mean Southern Oscillation Index and the yearly averaged storm severity in New South Wales. They generated a linear equation to estimate the extreme wave height on the basis of the SOI. Short and Trembanis (2004) observed decadal-scale oscillations in erosion-accretion and beach rotation on the order of 70 meters and 30 meters, respectively, at Narrabeen Beach, New South Wales. They suggested that these oscillations are closely linked to ENSO or the PDO.

Both mid-latitude and tropical storms affect the Australian Coast, as well as a class of storms that seems to be somewhat unique to the region, known as East Coast Cyclones (ECCs). ECCs are known to generate high winds and cause considerable damage along Australia's eastern coast during the late fall and early winter (Browning and Goodwin, 2013). These maritime storms are sub-tropical in origin, and may transform into mid-latitude cyclones or tropical cyclones depending upon conditions. Browning and Goodwin found that frequency of such storms is suppressed during El Niño years, although it does not necessarily increase during La Niña years. The complexity of these storms and the fairly poor understanding of their dynamics suggest that further study may provide valuable insight into episodic coastal change in eastern Australia.

ENSO has been shown to affect sea level and large-scale currents in western and southern Australia as well. Li and Clarke (2004) found that sea level was higher than normal during La Niña, causing eastward flow, and lower than normal during El Niña, inducing westward flow. The researchers highlighted the role of coastal orientation in influencing current dynamics. Although this oscillation affects continental shelf dynamics, it is somewhat unclear whether it has a significant effect on coastal response on a local scale.

New Zealand's northeast coast seems to be influenced by ENSO in much the same way as eastern Australia's, with La Niña occurrences enhancing beach erosion as a result of increased sea level and higher wave energy. However, de Lange (2001) notes that individual El Niño or La Niña phases are important to coastal management on scales of 5-8 years, but may be less significant in the long run than the Pacific Decadal Oscillation (PDO). The PDO appears to have shifted the system towards the La Niña extreme and increased coastal erosion between 1948 and 1976, with reduced erosion occurring in the subsequent 25 years as El Niño became more prevalent.

Other Global Impacts on Coastlines

ENSO is a global phenomenon that has the potential to affect coastlines worldwide. The strongest impacts that are well-documented in the literature occur along North America's West Coast, Australia's East Coast (and parts of New Zealand), equatorial areas of western South America (especially Ecuador and Peru), and mid-latitude regions of eastern South America (notably southern Brazil). However, coastal impacts connected to ENSO have been observed or at least hypothesized elsewhere. Black et al. (2008) found that the sediment budget on the monsoonal coast of Kerala, Southwest India, is maintained by a delicate balance of longshore and cross-shore transport. The authors speculate that such coasts could be very sensitive to ENSO fluctuations over time. El Niño events are associated with a rise in sea level and flooding in Kiribati. For example, a 1997 storm on Christmas Island caused a 0.5 m rise in sea level, which killed 40% of the island's coral (Office of the President, 2013). Although the impact of ENSO on Chinese coasts has not been well documented, Wang (1998) states that increased wave erosion occurs during El Niño events, resulting in beach and barrier retreat. Bers et al. (2013) found that even coastal Antarctica is influenced measurably by ENSO cycles, which appear to affect sea surface temperature, salinity, and suspended particulate levels.

A Brief Case Study from the Gulf of Mexico: Offshore Louisiana

Winter storms, including mid-latitude cyclones and cold front passages, are the dominant meteorological influence along the northern Gulf of Mexico between November and March. Aside from occasional tropical cyclones, winter storms are likely the greatest regional agents of coastal change, causing beach erosion and wetland loss in the region (Roberts et al., 1987; Stone and

Wang, 1999). During El Niño, storm frequency and precipitation typically increase over the Gulf of Mexico, at least between August and January, as a result of a shift in the jet stream (Munroe, 2013). It is reasonable to assume, therefore, that ENSO also has a measurable effect on coastal evolution, although this does not appear to have been well-studied.

The author and colleagues deployed instrumentation between November 24, 1998 and February 2, 1999 on the inner shelf of Louisiana, at depths of six to nine meters. The deployment occurred around Ship Shoal, a submerged Holocene barrier island, which, at the time was being considered as a source for nearshore sediment-nourishment (Stone and Xu, 1996). Unlike many of Louisiana's coastal environments, which are dominated by silt and mud, bed sediment in the study area was clean quartz sand with a mean grain diameter of 0.12-0.13 mm. Three bottom-mounted instrumentation systems were deployed, which included a combination of pressure transducers, current meters, optical backscatter sensors, and acoustic doppler velocimeters. The deployment occurred during the moderate La Niña episode that followed the intense El Niño event of 1997-98. Nonetheless, seven winter storms occurred during the deployment, separated by intervals of fair weather. The winter storms were divided into two basic types: migrating cyclones, which were characterized by a distinct low pressure cell and strong counterclockwise wind circulation; and arctic surges, which involved the passage of a cold front and an accompanying drop in air temperature and wind direction.

It was clear that winter storms caused significant increases in wave and current activity on the Louisiana inner shelf. Sediment transport was measured throughout the deployment, although the direction and magnitude were highly variable, not only as a result of different types of weather conditions, but also between storms, over the course of individual storms, and in some instances, between sites. However, several consistent results emerged. Sediment transport was offshore during winter storms and onshore during fair weather, as averaged over the course of the deployment. Perhaps not surprisingly, the strongest storms, which were migrating cyclones, caused much higher rates of sediment transport, causing brief but very high magnitude spikes in offshore sand movement. Onshore transport was sometimes evident during weaker storms, and was fairly consistent during fair weather, although generally at much lower rates than during strong storms. It appears, therefore, that winter storms may be important agents of offshore transport of fine sand during the winter, with onshore transport occurring much more gradually during fair weather (Pepper and Stone, 2002). Moreover, migrating cyclones were associated with the highest rates of sediment transport in the data record

(Pepper and Stone, 2004). El Niño conditions are likely to increase the frequency of winter storms in the region, and particularly of strong migrating cyclones, possibly shifting sand away, at least temporarily, from the coast. Although much of this sediment may eventually be recovered during fair weather, such a loss may be problematic, since the reduction in offshore sand may reduce shoaling behavior in waves, causing barrier islands to be more vulnerable to erosion. Hurricanes and tropical storms tend to receive most of the attention along the northern Gulf of Mexico, and perhaps justifiably so. Still, the role of ENSO in changing winter storm dynamics and influencing coastal processes in the Gulf is a topic that requires more study.

A Note on Coral Reefs

The focus of this chapter has been the physical responses of coastlines, particularly erosion and sedimentation, to forcing mechanisms such as wind, waves, and currents. However, the effect of ENSO on marine biology cannot be overlooked. Biology and geomorphology intersect very visibly at coral reefs, which also happen to be very sensitive to fluctuations in sea surface temperature and water level, two variables that are strongly affected by ENSO. ENSO has a strong impact on coral reef growth and health in many parts of the world, including the Western Indian Ocean (Zinke et al., 2005). Globally, coral bleaching was extensive during the 1997-98 ENSO episode with reefs in the Indian Ocean, the eastern and western Pacific, the Caribbean, and the Atlantic experiencing coral mortality rates as high as 99% in some locations (Goreau et al., 2000). Although the global loss of coral is a complex problem, driven by long-term climate change, pollution, sedimentation, and predation, the effect of ENSO is an important piece of the puzzle.

Conclusion

ENSO affects weather and ocean circulation worldwide, and is clearly an important consideration in coastal management. In many locations, however, its effects are negligible, or are at least nearly impossible to isolate. In other areas, ENSO has some measurable influence, but it is only one of many important factors. Finally, there are several coastlines where ENSO is the primary seasonal to interannual consideration for coastal response and management.

ENSO affects the genesis, evolution, and track of tropical cyclones in the Atlantic and the Pacific by modifying sea surface temperature, vertical wind shear, and overall wind circulation patterns. In the western Atlantic, including eastern North America and the Caribbean, severe hurricanes are several times more likely to make landfall during La Niña episodes than during El Niño. Some parts of the South Pacific, including Australia, may also be more susceptible during La Niña. Japan, Korea, and the central Pacific, on the other hand, are more likely to experience tropical cyclone landfalls during El Niño. However, because of the somewhat low frequency of tropical cyclone landfalls along any particular coastline (regardless of ENSO circulation pattern) it is difficult to offer specific recommendations for coastal zone management in this context.

The most predictable and easily-addressed effect of ENSO on coastlines occurs where it modifies the intensity, frequency, and track of winter storms, such as mid-latitude cyclones and frontal passages. Areas that fit this profile include the West Coast of North America, the Equatorial Pacific Coast of South America, and the Atlantic Coasts of Brazil and Argentina, where El Niño phases increase the magnitude of storm waves and currents and cause enhanced erosion and land loss. Similar effects are observed in New Zealand and along the East Coast of Australia during La Niña episodes. There are several other areas where ENSO might play an important role in coastal evolution that require more study, including the Indian Ocean and the Gulf of Mexico.

Even in areas where the meteorological and hydrodynamic effects of ENSO are fairly well documented, it is clear that it affects coastal types quite differently. On sandy beaches, ENSO may cause a dramatic but fairly short-term increase in erosion. In many cases, these beaches fully recover within a few months or years. Moreover, on many of these coasts, ENSO provides pulses of sediment that are vital to long-term supply. Averaged over a period of decades or longer, the impact of ENSO on beaches may be analogous to its role in the Earth's climate system. ENSO may represent a temporary oscillation in shoreline position embedded within long-term trends of relative sea-level change and regional sediment supply. On the other hand, coastal managers must be responsive to short-term changes in addition to long-term trends, and ENSO can generate erosion hotspots that may persist for many years, particularly when coastal infrastructure such as groins and jetties interrupt longshore sediment transport. On cliffed coasts, ENSO may lead to dramatic rates of erosion and retreat that exceed the long-term annual average by orders of magnitude. Cliffs, of course, cannot recover during non-ENSO periods, so these environments present a difficult challenge to coastal managers.

Finally, coral reefs are among the most threatened coastal features on Earth, owing to a combination of anthropogenic, physical, and biological factors. Because of their sensitivity to temperature, reef building corals are extremely vulnerable to long-term climate change. However, ENSO episodes are also devastating over a shorter time-scale, causing increases in sea surface temperature and water level that can trigger mortality rates of more than 90%. It is difficult to know how to address this problem from a management perspective; the stark reality is that we may be headed towards a world without living coral reefs.

Research continues to increase our understanding of ENSO, its variations, and its links to other atmospheric teleconnections and long-term climate change. It has been three decades since the intense 1982-83 El Niño episode wreaked havoc around the planet, catching some of the scientific community and most of the public off-guard; since then, our level of understanding and preparedness has grown immensely. The National Weather Service now maintains a website devoted to ENSO, and issues predictions weeks or months in advance, and coasts that once lay fairly unprotected are now managed much more effectively. Still, ENSO remains an critical agent of change along many coastlines that will continue to pose a challenge to coastal managers for the foreseeable future.

References

Adams, P.N., Inman, D.L., & Graham, N.E. (2008). Southern California deep-water wave climate: Characterization and application to coastal processes. *Journal of Coastal Research*, 24(4), 1022-1035.

Allan, J.C. & Komar, P.D. (2006). Climate Controls on US West Coast Erosion Processes. *Journal of Coastal Research*, 22 (3), 511–529.

Allan, J.C., Komar, P.D., & Priest, G.R. (2003). Shoreline variability on the high-energy Oregon coast and its usefulness in erosion-hazard assessments. *Journal of Coastal Research,* SI 38, 83-105.

Andrews, E.D., Antweiler, R.C., Neiman, P.J., & Ralph, F.M. (2004). Influence of ENSO on Flood Frequency along the California Coast. *Journal of Climate,* 17, 337-348.

Ashok, K., Behera, S.K., Rao, S.A., Weng, H., & Yamagata, T. (2007). El Niño Modoki and its possible teleconnections. *Journal of Geophysical Research,* 112 (C11007), doi:10.1029/2006JC003798.

Barnard, P. L., J. Allan, J. E. Hansen, G. M., Kaminsky, P. Ruggiero, & A. Doria (2011) The impact of the 2009–10 El Niño Modoki on U.S. West Coast beaches, *Geophysical Research Letters*, 38, L13604.

Bers, A.V., Fernando, M., Schloss, I.R., & Abele, D. (2013) Analysis of trends and sudden changes in long-term environmental data from King George Island (Antarctica): relationships between global climatic oscillations and local system response. *Climatic Change*, 116 (3-4), 789-803.

Bjerknes, J. (1969). Atmospheric Teleconnections from the Equatorial Pacific. *Monthly Weather Review* 97 (3), 163-172.

Black, K.P., Kurian, N.P., Mathew, J., & Baba, M. (2008). Open coast monsoonal beach dynamics. *Journal of Coastal Research*, 24 (1), 1-12.

Bove, M.C., O'Brien, J.J., Elsner, J.B., Landsea, C.W., & Niu, X. (1998). Effect of El Niño on U.S. Landfalling Hurricanes, Revisited. *Bulletin of the American Meteorological Society,* 79 (11), 2477-2482.

Brooke, B. (2013). Beach Erosion. Oz Coasts: Australian Online Coastal Information. Geoscience Australia. Accessed on: *http://www.ozcoasts.gov.au/indicators/beach_erosion.jsp, on* 10/08/2013.

Browning, S.A. & Goodwin, I.D. (2013). Large-Scale Influences on the Evolution of Winter Subtropical Maritime Cyclones Affecting Australia's East Coast. *Monthly Weather Review*, 141, 2416-2431.

Chan, J. & Zhou, W. (2005). PDO, ENSO and the early summer monsoon rainfall over south China. *Geophysical Research Letters,* 32 (8), L08810, doi:10.1029/2004GL022015.

Chen, W.Y. (1982). Fluctuations in Northern Hemisphere 700 mb Height Field Associated with the Southern Oscillation. *Monthly Weather Review*, 110: 808-823.

Chu, P.-S. (2004). ENSO and tropical cyclone activity. In: R.J. Murnane and K.-B. Liu (Eds.), *Hurricanes and Typhoons: Past, Present, and Potential (pp.* 297-332). New York: Columbia University Press.

Climate Prediction Center (2013). Cold and Warm Episodes by Season. NOAA National Weather Service: Accessed at *http://www.cpc.ncep.noaa.gov/products/analysis_monitoring/ensostuff/ensoyears.shtml*, on 09/10/2013.

Climate Prediction Center (2013). Monthly Atmospheric and SST Indices. NOAA National Weather Service: Accessed at *http://www.cpc.ncep.noaa.gov/data/indices/soi.3m.txt*, on 09/10/2013.

Cobb, K.M., Westphal, N., Sayani, H.R., Watson, J.T., Di Lorenzi, E., Cheng, H., Edwards, R.L., & Charles, C.D. (2013). Highly Variable El Niño-Southern Oscillation Throughout the Holocene. *Science*, 339 (6115), 67-70.

Dahlman, L. (2009). Climate Variability: Ocean Niño Index. Climate.gov. Accessed at: *http://www.climate.gov/news-features/understanding-climate /climate- variability-oceanic-ni%C3%B1o-index*, on 09/17/2013.

de Lange, W.P. (2001). Interdecadal Pacific Oscillation (IPO): A mechanism for forcing decadal scale coastal change on the northeast coast of New Zealand? *Journal of Coastal Research*, SI 34, 657-664.

de Souza, M.C. & Angulo, R.J. (2003). Decadal and Inter-Annual Variations of Shoreline and Beach Volumes in Itapoá (Santa Catarina, Brazil). *Journal of Coastal Research,* SI 35, 202-208.

Earth Systems Research Laboratory: Physical Sciences Division (2013a). Multivariate ENSO Index: MEI. NOAA: Accessed at: *http://www. esrl.noaa.gov/psd/enso/mei/, on* 09/17/2013.

Earth Systems Research Laboratory: Physical Sciences Division (2013b). What's happening in the atmosphere during El Niño. NOAA: Accessed at: *http://www.esrl.noaa.gov/psd/enso//enso.description.html*, on 10/24/2013.

Esteves, L .S., Williams, J. J., & Dillenburg, S.R. (2006). Seasonal and interannual influences on the patterns of shoreline changes in Rio Grande do Sul, southern Brazil. *Journal of Coastal Research*, 22(5), 1076-1093.

Etayo-Cadavid, M.F., Andrus, C.F.T., Jones, K.B., Hodgins, G.W.L., Sandweiss, D.H., Uceda-Castillo, S., & Quilter, J. (2013). Marine radiocarbon reservoir age variation in Donax obesulus shells from northern Peru: Late Holocene evidence for extended El Niño. *Geology*, 41(5), 599-602.

Fairbridge, R.W. (1998). El Niño not Unpredictable as Coastal Hazard: Luni-Solar Synergy. *Journal of Coastal Research*, SI 26, 52-63.

Faraco, K.R., de Castilhos, J.A., & Horn Filho, N.O. (2006). Morphodynamic Aspects and El Niño Oscillations in Ingleses Beach, Santa Catarina Island, Southern Brazil. *Journal of Coastal Research,* SI 39, 656-659.

Federici, P.R, & Rodolfi, G. (2001). Rapid shoreline retreat along the Esmeraldas coast, Ecuador: natural and man-induced processes. *Journal of Coastal Conservation,* 7, 163-170.

Fedorov, A.V. & Philander, S.G. (2000). Is El Niño changing? *Science*, 288, 1997-2002.

Giese, B.S., Compo, G.P., Slowey, N.C., Sardeshmukh, P.D., Carton, J.A., Ray, S., & Whitaker, J.S. (2010). The 1918/19 El Niño. *Bulletin of the American Meteorological Society,* 91, 177-183.

Glantz, M.H. (2000). *Currents of Change: Impacts of El Niño and La Niña on Climate and Society* (2e.). Cambridge, U.K.: Cambridge University Press.

Goreau, T., McClanahan, T., Hayes, R., & Strong, A. (2000). Conservation of Coral Reefs after the 1998 Global Bleaching Event. *Conservation Biology,* 14 (1), 5-15.

Graham, N.E. & White, W.B. (1988). The El Niño Cycle: A Natural Oscillator of the Pacific Ocean-Atmosphere System. *Science*, 240, 1293-1302.

Griggs, G., Patsch, K., & Savoy, L. (Eds.) (2005). *Living with the Changing California Coast.* Berkeley, CA: University of California Press.

Hapke, C.J., Reid, D., & Richmond, B. (2009). Rates and Trends of Coastal Change in California and the Regional Behavior of the Beach and Cliff System. *Journal of Coastal Research*, 25(3), 603-615.

Hudson, P.F. (2003). The Influence of the El Niño Southern Oscillation on suspended sediment load variability in a seasonally humid tropical setting: Pinuco Basin, Mexico. *Geografiska Annaler,* 85 A (3-4), 263-275.

Inman, D.L. & Jenkins, S.A. (1999). Climate Change and the Episodicity of Sediment Flux of Small California Rivers. *Journal of Geology,* 107, 251-270.

Johnson, Nathaniel (2013). How many ENSO flavors can we distinguish*?* *Journal of Climate*. 26 (12), 4816-4827.

Komar, P.D. (1998). El Niño and coastal erosion in the Pacific Northwest. *Oregon Geology*, 60 (3), 57-64.

Kug, J.-S., Jin, F.-F., & An, S.-I. (2009). Two types of El Niño Events: Warm Pool El Niño and Cold Tongue El Niño. *Journal of Climate*. 22, 1499-1514.

Lamour, M.R. & Soares, C.R. (2003). Sedimentary Variation and Volumetric Balance in Four Beaches in North Littoral of Santa Catarina State, Brazil in a La Niña/El Niño Period. *Journal of Coastal Research,* SI 35, 216-220.

Li, J. & Clarke, A.J. (2004). Coastline Direction, Interannual Flow, and the Strong El Niño Currents along Australia's Nearly Zonal Southern Coast. *Journal of Physical Oceanography*, 34: 2373-2381.

Maia, L.P.; Friere, G.S.S., & Lacerda, L.D. (2005). Accelerated dune migration and aeolian transport during El Nifio events along the NE Brazilian coast. *Journal of Coastal Research*, 21(6), 1121-1126.

Martinez, J.O., Gonzalez, J.L.; Pilkey, O.H., & Neal, W.J. (1994). Tropical barrier islands of Colombia's Pacific coast. *Journal of Coastal Research*, 11(2), 432-453.

Masters, P.M. (2006). Holocene sand beaches of southern California: ENSO forcing and coastal processes on millennial scales. *Palaeogeography, Palaeoclimatolgy, Palaeoecology*, 232(1), 73-95.

McPhaden, M.J., Zebiak, S.E., & Glantz, M.H. (2006). ENSO as an Integrating Concept in Earth Science. *Science*, 314, 1740-1745.

Munroe, R., Crawford, T., & Curtis, S. (2013). Geospatial Analysis of Space–Time Patterning of ENSO Forced Daily Precipitation Distributions in the Gulf of Mexico. *The Professional Geographer*, DOI:10.1080/ 00330124.2013.765291.

National Oceanic and Atmospheric Administration (2013a). El Niño Theme Page. Accessed at: *http://www.pmel.noaa.gov/tao/elNiño/el-Niño-story.html, on 11/04/2013.*

National Oceanic and Atmospheric Administration (2013b). NOAA's El Niño Page. Accessed at*: http://www.elnino.noaa.gov/*, on 11/05/2013.

National Weather Service (2012). Weather Impacts of ENSO. JetStream - Online School for Weather. Accessed at: *http://www.srh.noaa.gov/ jetstream/tropics/enso_impacts.htm, on 10/24/2013.*

Office of the President of the Republic of Kiribati (2013). Kiribati Climate Change: Coastal Erosion. Accessed at: *http://www.climate. gov.ki/category/effects/coastal-erosion/,* on 10/08/2013.

Palumbo, A. (2001). Long-Term Forecasting of the Extreme El-Niño Events. *Journal of Coastal Research*, 17 (2), 394- 397.

Pepper, D.A. & Stone, G.W. (2002). Atmospheric forcing of fine-sand transport on a low-energy inner shelf: south-central Louisiana, USA. *Geo-Marine Letters*, 22, 33-41.

Pepper, D.A. & Stone, G.W. (2004). Hydrodynamic and sedimentary responses to two contrasting winter storms on the inner shelf of the northern Gulf of Mexico. *Marine Geology*, 210, 43-62.

Philander, S. G. H. (1985). El Niño and La Niña, *Journal of Atmospheric Science,* 42, 2652-2662.

Phinn, S.R. & Hastings, P.A. (1992). Southern oscillation influences on the wave climate of south-eastern Australia. *Journal of Coastal Research*, 8 (3), 579-592.

Phinn, S.R., & Hastings, P.A. (1995). Southern oscillation influences on the Gold Coast's summer wave climate, *Journal of Coastal Research*, 11 (3), 946-958.

Quan, S., Kvitek, R., Smith, D.P., & Griggs, G.B. (2013) Using Vessel-Based LIDAR to Quantify Coastal Erosion during El Niño and Inter-El Niño Periods in Monterey Bay, California. *Journal of Coastal Research*, 29 (3), 555-565.

Ramesh, N. & Murtugudde, R. (2013). All flavors of El Niño have similar subsurface origins. *Nature Climate Change* 3, 42–46.

Revell, D.L., Komar, P.D., & Sallenger, A.H., Jr. (2002). An Application of LIDAR to Analyses of El Niño Erosion in the Netarts Littoral Cell, Oregon. *Journal of Coastal Research*, 18 (4), 792-801.

Revell, D.L., Dugan, J.E., & Hubbard, D.M. (2011). Physical and Ecological Responses of Sandy Beaches to the 1997-98 El Niño. *Journal of Coastal Research,* 27(4), 718-730.

Roberts, H.H., Huh, O.K., Hsu, S.A., Rouse, Jr., L.J. & Rickman D. (1987). Impact of Cold-Front Passages on Geomorphic Evolution and Sediment Dynamics of the Complex Louisiana Coast. *Proceedings of Coastal Sediments '87,* 2, 1950-1963.

Rosenthal, Y. & Broccoli, A.J. (2004). In Search of Paleo-ENSO. *Science*, 304, 219-220.

Sandweiss, D. H., McInnis, H., Burger, R.L., Cano, A., Ojeda, B., Paredes, R., Sandweiss, M., & Glascock, M. (1998). Quebrada Jaguay: Early South American Maritime Adaptations. *Science*, 281, 1830-1832.

ScienceDaily (2000). Historic Records Reveal Links Between El Niño, Coastal Erosion, And Shifting Sands Of Beaches In Central California. Accessed at: *http://www.sciencedaily.com/releases/2000/12/00122708 2034.htm*, on 10/25/2013.

Short, A.D. & Trembanis, A.C. (2004). Decadal Scale Patterns in Beach Oscillation and Rotation Narrabeen Beach, Australia. *Journal of Coastal Research,* 20 (2), 523-532.

Stevenson, S., Fox-Kemper, B., Jochum, M., Neale, R., Deser, C., & Meehl, G. (2012). Will There Be a Significant Change to El Niño in the Twenty-First Century?. *Journal of Climate*, 25, 2129–2145.

Stone, G.W. & Wang, P. (1999). The Importance of Cyclogenesis on the Short-Term Evolution of Gulf Coast Barriers. *Transactions of the Gulf Coast Association Geological Societies,* 49, 478-487.

Stone, G.W. & Xu, J.P. (1996). Wave climate modeling and evaluation relative to sand mining on Ship Shoal, offshore Louisiana, for coastal and barrier island restoration. *Coastal Morphodynamics Laboratory Final Report*, Louisiana State University.

Storlazzi, C.D. & Griggs, G.B. (1998). Influence of El Niño-Southern Oscillation (ENSO) Events on the Coastline of Central California. *Journal of Coastal Research*, SI 26, 146-153.

Storlazzi, C.D. & Griggs, G.B. (2000). Influence of El Niño-Southern Oscillation (ENSO) events on the evolution of central California's shoreline. *GSA Bulletin,* 112 (2), 236-249.

Storlazzi, C.D., Willis, C.M. & Griggs, G.B. (2000). Comparative impacts of the 1982-83 and 1997-98 El Niño winters on the central California coast. *Journal of Coastal Research,* 16 (4), 1022-1036.

Sumesh, K.G., & Ramesh Kumar, M.R. (2013). Tropical Cyclones over north Indian Ocean during El Niño Modoki years. *Natural Hazards*, 68, 1057-1074.

Thompson, C., Ready, S., & Zheng, X. (1992). Tropical Cyclones in the Southwest Pacific, November 1979 to May, 1989. New Zealand Meteorological Service. Accessed at *http://docs.niwa.co.nz/library/public/0477073468.pdf, on 09/08/2013.*

Trenberth, K.E. (1997). The Definition of El Niño. *Bulletin of the American Meteorological Society*, 78 (12), 2771-2777.

University of Illinois (1999). El Niño Surface Temperatures. WW2010 Project. Accessed at *http://ww2010.atmos.uiuc.edu/(Gh)/guides/mtr/eln/sst.rxml,* on 09/07/2013.

USGS (2013). Coastal Erosion Along the U.S. West Coast During the 1997-98 El Niño: Expectations and Observations. Center for Coastal Geology. Accessed at *http://coastal.er.usgs.gov/lidar/AGU_fall98/, on 10/08/2013.*

USGS (2013). Coastal Change Hazards: Hurricanes and Extreme Storms. St. Petersburg Coastal and Marine Science Center. Accessed at *http://coastal.er.usgs.gov/hurricanes/elNiño/, on 10/08/2013.*

Walker, G.T. (1924). Correlation in Seasonal Variations of Weather, IX. *Memoirs of the India Meteorological Department,* Vol. XXIV.

Walker, G.T., & Bliss, E.W. (1932). *Memoirs of the Royal Meteorological Society*. IV (36), 53-84.

Wang, Y. (1998). Sea-Level Changes, Human Impacts and Coastal Responses in China. *Journal of Coastal Research*, 14(1), 31-36.

Wang, H.-J., Zhang, R.-H., Cole, J., & Chavez, F. (1999). El Niño and the related phenomenon Southern Oscillation (ENSO): The largest signal *in interannual climate variation. Proceedings of the National Academy of Sciences*, 96, 11071-11072.

Weng, H., Ashok, K., Behera, S.K., Rao, S.A., & Yamagata, T. (2007). Impacts of recent El Niño Modoki on dry/wet conditions in the Pacific rim during boreal summer. *Climate Dynamics*, 29, 113-129.

Wolter, K., & Timlin, M.S. (1993). Monitoring ENSO in COADS with a Seasonally Adjusted Principal Component Analysis (pp. 52-57). *Proceedings of the 17th Climate Diagnostics Workshop.* Norman, OK: NOAA/NMC/CAC, NSSL, Oklahoma Climate Survey, CIMMS and the School of Meteorology, University of Oklahoma.

Wolter, K., & Timlin, M.S. (1998). Measuring the strength of ENSO events: How does 1997/98 rank? *Weather*, 53(8): 315-323.

Wyrtki, K. (1973). Teleconnections in the Equatorial Pacific Ocean. *Science*, 180, 66-68.

Yoon, J. & Yeh, S. -W. (2010). Influence of the Pacific Decadal Oscillation on the Relationship between El Niño and the Northeast Asian Summer Monsoon. *Journal of Climate*, 23, 4525-4537.

You, Z.-.J, & Lord, D. (2008). Influence of the El Niño-Southern Oscillation on NSW coastal storm severity. Journal of Coastal Research, 24 (2B), 203-207.

Zhang, W., Graf, H.-F., Leung, Y., & Herzog, M. (2012). Different El Niño Types and Tropical Cyclone Landfall in East Asia. *Journal of Climate,* 25, 6510-6523.

Zinke, J, Pfeiffer, M., Timm 0., Dullo W.-C., & Davies, G. R (2005). Atmosphere-ocean dynamics in the Western Indian Ocean recorded in corals. *Philosophical Transactions of the Royal Society*, 363, 121-142.

In: Beaches
Editor: Margaret Cessa
ISBN: 978-1-63117-239-7

Chapter 2

COMMENTARY: MACROFAUNAL KEY SPECIES AS USEFUL TOOLS IN MANAGEMENT AND ASSESSMENT PRACTICES ON SANDY BEACHES

Sílvia C. Gonçalves[1,2,#] and João C. Marques[1,*]
[1]IMAR – Marine and Environmental Research Centre, Department of Life Sciences, Faculty of Sciences and Technology, University of Coimbra, Largo Marquês de Pombal, Coimbra, Portugal
[2]GIRM - Marine Resources Research Group, School of Tourism and Maritime Technology, Polytechnic Institute of Leiria, Campus 4, Santuário Na. Sra. dos Remédios, Peniche, Portugal

ABSTRACT

Sandy beaches are the most abundant coastal systems worldwide, representing approximately ¾ of the worlds' coastlines. In the last decades, these sensitive and dynamic ecosystems have been a target of an increasing human pressure, associated with an unbridled urbanization of coastal areas, caused by the high human population growth, but also associated with recreational activities and tourism. Together with global climatic changes and coastal erosion, these processes lead to detrimental impacts and an environmental quality decline of sandy beach ecosystems,

[#] E-mail: scgoncalves@ipleiria.pt; Tel: +351 262 783 607; Fax: +351 262 783 088.
[*] E-mail: jcmimar@ci.uc.pt.

imposing an increasing need for adequate assessment and management practices. In the majority of sandy beaches, the macrofauna is a key element in ecosystem functioning, occupying a central position in the food webs. Some species exhibit abundant populations, representing the bulk of the macrofaunal communities and functioning as key species in these conditions. In this commentary we discuss the relevance and utility of macrofaunal key species as tools in beaches integrity assessment, planning of management practices and restoration projects. We start by exposing the reasons behind the selection and use of these organisms as potential tools in assessment and management practices of beaches. Several and alternative methodologies, at the population and at the community ecological levels of organization, are afterwards discussed and their advantages and drawbacks evaluated. Finally, the most relevant key species in worldwide sandy beaches, known in the literature, and with potential as assessment and management tools, are identified. Since several manmade disturbances (e.g., intense human and mechanical trampling, beach grooming with heavy mechanical equipment), and even natural disturbances (e.g., abnormal temperature regimes, frequency and intensity of storms) may cause variations in key species bioecology, with negative impacts perpetuated to the higher ecological levels, monitoring, assessment and management actions focused on this indicator species hold large potential in the fast changing scenario of nowadays, and should be encouraged.

1. INTRODUCTION

Approximately ¾ of the world's coastlines are sandy beaches. Coastal areas have always been focal places for the establishment of Humans, and the economical and aesthetical values, namely regarding the exploitation of natural resources (living and non-living), relaxation and leisure possibilities, have always attracted human populations to living near the beach.

In the last decades, these sensitive and dynamic ecosystems have been a target of a growing human pressure. Approximately 50% to 70% of the population lives within 60 km of the coast, a proportion that is actually increasing (Caffyn et al., 2002). Since the 1950s, an enormous growth in beach tourism has occurred (Gormsen, 1997; Caffyn et al., 2002), taking more and more tourists to these coastal systems every year. Unfortunately, mass tourism in coastal areas is nowadays an undeniable reality. This development has been only possible due to an escalating expansion of touristic and other recreational associated facilities, such as hotels, resorts, jetties, service and parking areas, just to name a few. All these human activities introduced detrimental impacts and led to an environmental quality decline of the beach environments (e.g., Weslawski et al., 2000; Lercari and Defeo, 2003; Davenport and Davenport, 2006). Also, environmental problems related with climate change and sea level rise exacerbate this environmental quality decline, conducting for instance to beach erosion and narrowing (McLachlan and Brown, 2006; Doney et al., 2012). Thus, beach assessment, management and protection are nowadays extremely urgent. Although a very large effort, and several important contributions and advances, have been achieved in the last 15 to 20 years on the study and understanding of the biological and ecological components of sandy beaches, little is known about the responses of these ecosystems to their distinct stressors and pressures. It is therefore not surprising that beach assessment and management strategies are still very poorly developed and beach conservation is still difficult to perceive in the horizon (but see for instance Nordstrom, 2008; McLachlan et al., 2013).

Crustaceans, molluscs and polychaetes are often the most common inhabitants of the macrofaunal communities of sandy beaches, although several other groups may also be present (McLachlan and Brown, 2006). In the vast majority of these systems, macrofauna is a fundamental element in ecosystem structure and function. Their presence in adequate numbers ensures the equilibrium of these ecological systems and allows sandy beaches to operate normally as ecosystems and perform several important functions and services (e.g., decomposing centres for coastal debris, biological filters in nutrient recycling, foraging and breeding areas for several vertebrates). Macrofauna contributes for wrack consumption and degradation and, simultaneously, serves as prey for several top predators such as fishes or shorebirds, occupying a central position in the beach food webs. Also, some species, like for instance sand hoppers in temperate beaches (e.g., Gonçalves et al., 2009) or ghost crabs on tropical and subtropical ones (e.g., Neves and Bemvenuti, 2006), exhibit abundant populations and may sometimes represent the bulk of the macrofaunal communities (see for instance Gonçalves et al., 2009; Gonçalves, 2012).

In this commentary we discuss the relevance and the utility of macrofaunal key species as tools in beaches integrity assessment, planning of management practices and restoration projects. We start by exposing the reasons behind the selection and use of these organisms as potential tools in assessment and management practices of beaches. Several and alternative methodologies, at the population and at the community ecological levels of organization, are afterwards discussed and their advantages and drawbacks evaluated. Finally, the most relevant key species in worldwide sandy beaches, known in the literature, and with potential as assessment and management tools, are identified.

2. Why Should We Use Macrofaunal Key Species in Assessment and Management Practices?

In coastal systems, abundant invertebrate populations are often the driving forces of the community and, by shaping both the community's structure and function, have a relevant role on the equilibrium of the ecosystem (see for instance Gonçalves et al., 2009; Gonçalves and Marques, 2011). This is a clear and simple definition of a *key species*. Removal or depletion of a key species, as well as strong abnormal population growths, translates into profound effects in the structure of the community and on the ecosystems equilibrium. On the other hand, and because this equilibrium is dynamic, changes in the ecosystem will also be reflected in key species populations (figure 1).

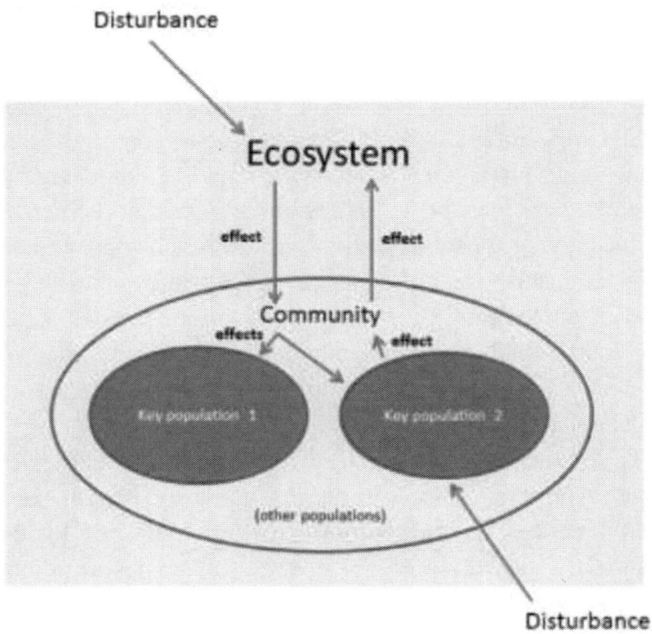

Figure 1. Schematic representation of the dynamic equilibrium between an ecosystem and its key species, when subjected to a disturbance.

Regarding management, key species are important *indicators of sustainability* (Scapini and Morgan, 2002), but may also function as *indicators of environmental changes* in the complex network of the beach ecosystem. Of course, from an ecological management point of view, it is also extremely important that the key species is simultaneously widespread and nevertheless fairly sensitive to the human pressures on the beach.

As previously mentioned, *macrofauna* is a fundamental and central component on beach ecosystems, essential for an adequate performance of the ecological functions of these ocean/land interface systems. Also, beaches are known to be dynamic, but also sensitive, ecosystems that change rapidly *per se*, especially in open oceanic conditions. In urbanized beaches the intensity and frequency of these changes is aggravated by the disturbances derived from human activities. Therefore, a good and potentially effective approach to use in beach management and assessment practices would be to couple the virtues of macrofauna and key species, by selecting macrofaunal key species and using them as practical tools. In the following section, distinct methodologies to do so will be presented.

3. How Can They Be Used?

Due to the structural and functional duality that characterize the role of key species in the ecosystem, macrofaunal key species may be used as ecological indicators of ecological quality condition, sustainability or of environmental changes (natural or induced by specific human impacts on the beach) using two distinct approaches: the population and the community.

3.1. The Population Approach

A rapid changing environment is one of the most characteristic features of sandy beaches. This occurs mainly because beaches are often physically controlled, and is more pronounced on exposed conditions. Local populations respond to each change induced in the system through adaptation and, according to Hazlett (1988), behavioural plasticity is the most immediate and common adaptation to environmental change for intertidal organisms. Therefore, studying a macrofaunal key species at the population level is an obvious and advantageous approach for the assessment of the impacts of natural or human induced environmental changes on beach ecosystems.

One of the tools that may be used in management and assessment practices of the beach ecosystem at the population level is *bioecology.* This approach has been proposed by some authors (e.g., Marques and Anastácio, 2002; Gonçalves 2012; Gonçalves et al., 2013), namely as a monitoring and assessment tool of the sandy beach ecological quality condition (see Gonçalves et al., 2013 for more details). Bioecological features such as population dynamics (densities, mortality, reproduction and recruitment), reproductive biology (extension of the reproductive period, fecundity and voltinism) and secondary production estimates (growth and elimination productions, average annual biomass and population turnovers) are particularly relevant and useful. For instance, oscillations in the reproductive strategies or in the population dynamics of a given species may reflect the occurrence of a specific environmental change.

Knowing and understanding the bioecology of macrofaunal key populations of undisturbed beaches or of beaches with a small degree of human disturbance, provides relevant information about the adaptability of populations to natural sources of variation (e.g., food resources availability, sand grain size, competition). Also, comparative studies of distinct populations on local and macro scale (macroecology) scenarios allow the detection of variations on population bioecology resulting from the adaptability of the populations to the local scale environments or even to large-scale latitudinal variations, as proposed by Gonçalves et al. (2013). All together, the construction of solid baseline knowledge on key populations bioecology is a very useful tool since it is much easier to detect variations in population biology derived from anthropogenic pressures, if natural sources of variation are known *a priori* and may therefore be properly discarded. Such a vast bioecological knowledge about a key species will undoubtedly function as a reference knowledge tool for further ecosystem integrity assessments (Gonçalves, 2012), or for environmental monitoring and ecological quality assessments in beaches with similar physical conditions (Gonçalves et al., 2013), after major environmental disturbances take place, like for instance caused by human activities.

Significant decreases in population densities of a known macrofaunal key population will produce serious effects in population ecology. Depending on the nature of the disturbance, changes in population dynamics and in the reproductive strategies may occur, as a response to increased mortality rates. Also, if the disturbance is reflected in population bioecology, secondary production attributes and population turnovers will also denunciate the disturbance, deviating from the reference attributes (figure 2).

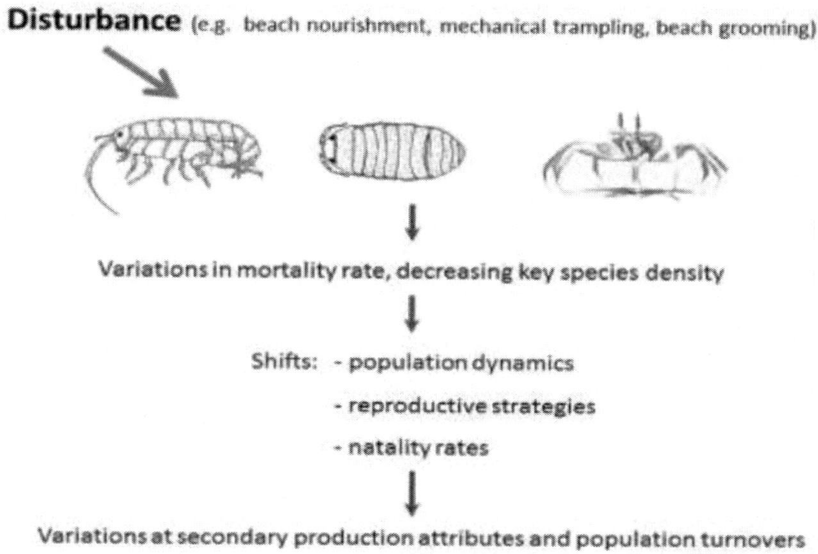

Figure 2. Schematic representation of the potential consequences of a disturbance, namely derived from human activities, on the bioecology of macrofaunal key species of sandy beaches.

Several manmade disturbances may conduct such a chain of events on sandy beach macrofaunal populations. For example, increases in touristic and recreational activities on beaches produce several kinds of impacts that may increase the mortality rates of key populations. Intense human trampling and also mechanical trampling by off-road vehicles, used for instance in recreational activities, can damage the animals, by destroying their burrows or even fatally crush them (see for instance Węslawski et al., 2000; Veloso et al., 2006; Schlacher et al., 2008). Beach grooming with heavy mechanical equipment on a regular basis can also result in density decreases of key populations, especially if they are wrack-associated fauna (e.g., Dugan et al., 2003 and as reviewed in Colombini and Chelazzi, 2003 and in Defeo et al., 2009). The constructions of touristic, recreational and some urban associated facilities are often responsible for coastal dune degradation and destruction. This kind of impact disturbs both the dune as the beach biota, especially supralittoral organisms (see for instance Barros, 2001; Lucrezi et al., 2008). Besides these examples of human impacts, mostly related with the increasing demand for sandy beaches as touristic and recreational targets, several others with potential to induce changes of macrofaunal key populations could be mentioned. These may include, for instance, climatic changes (e.g., abnormal temperatures, changes in storm occurrence and intensity), beach erosion, beach nourishment (e.g., Jones et al., 2008; Schlacher et al., 2012) and human harvesting (as reviewed in Defeo et al., 2009).

The effectiveness of using bioecology in assessment, monitoring and management practices might be dependent of the species life history traits. According to Gonçalves et al. (2013), monitoring species with slow life histories (e.g., extended lifespan, delayed reproduction, low population turnovers) allows detecting the occurrence of any given environmental disturbance more quickly and more easily, namely the ones derived from human activities. By contrast, due to their greater colonizing abilities, monitoring species with fast life histories might be more useful in environmental management programs of already disturbed sandy beaches, allowing to asses more quickly the effectiveness of a specific restoration or mitigation measure (Gonçalves et al., 2013). Notwithstanding, it should not be forgotten that the use of population bioecology *per si* may be a longstanding process, often focused on long term responses and at expense of heavy field and laboratory programs. Identifying population key variables or population key attributes and monitoring their progress using accurate sampling designs and data analysis may however attenuate these limitations.

Another important tool at the population level is the construction of *population dynamics models*, capable of comprehensively and correctly simulate the dynamics of a given key population. It is in fact a very powerful tool because it allows the ecological manager to simulate multiple scenarios that might reflect potential changes in environmental conditions, namely induced by human activities. Moreover, these models may also be used to understand or to mitigate the impacts that environmental changes, driven for instance by human activities, induce upon the modelled population (Anastácio et al., 2003). The simulation of distinct scenarios forecasts the populations' responses to the environmental change introduced in the modelling process, but also shows which parameters will predictably change and to what extent. By building a population dynamics model of a key species, indicator of a given biological community, the response of the system to a given change may be anticipated (Marques and Anastácio, 2002).

Modelling the population dynamics of macrofaunal key species, the following scenarios may, for instance, be simulated: (1) significant decreases in food availability caused by intense grooming activities; (2) significant increases or decreases in temperature caused by abnormal temperature regimes; (3) significant increases in human trampling caused by increased numbers of tourists or other beach users. These simulations will reveal the most probable responses of the population and, using these responses, changes in the communities may be anticipated and preventive management or mitigation actions implemented. However, accurate and reliable modelling outputs are only achievable provided that the model is correctly developed, calibrated and ideally validated. The completion of these steps of the model construction process requires a lot of time, solid baseline bioecological knowledge, with field datasets of natural populations with distinct origin, and off course great expertise. Furthermore, despite of the complex and time consuming process of model development, once obtained, a population dynamics model of a key species must be seen as a very quick and powerful simulation tool for environmental change assessment in sandy beaches, especially for changes derived from human activities (Gonçalves et al., 2013).

3.2. The Community Approach

According to several studies (e. g. Jones et al., 2008; Fanini et al., 2009; Schlacher et al., 2012), sandy beach macrobenthic communities can respond relatively rapid to natural and anthropogenic disturbances. Therefore, *community structural attributes* that are easily visualised, like for instance abundance of common species (especially key species), species number and diversity indexes are a good approach that can be used to monitor short and long term responses. This approach has been often used by sandy beach ecologists to assess for example the effects of some specific human activities on macrofaunal communities (see for instance Schlacher et al., 2008; Fanini et al., 2009; Leewis et al., 2012; Bessa et al., 2013). However, and as mentioned for the population approach, previous information upon what to expect of the structural attributes on undisturbed (or as less disturbed possible) communities are needed, if potential drivers of community changes are to be identified. Once obtained, these reference structural attributes may function as a basis for further comparison in ecosystem integrity assessment studies, in management planning and assessment, or even in ecosystem restoration programs. Moreover, determining the ecological condition of a given macrofaunal community is a fundamental practice on ecosystem assessment programs.

Ecosystem structure and function are deeply linked and cannot be dissociated from each other. In fact, some authors claim that both types of attributes, structural and functional, should be employed in ecosystem integrity assessment, especially after implementing restoration programs (Mayer and Galatowitsch, 2001; Mayer et al., 2004). *Community functional attributes*, like for instance trophic structure (functional groups), secondary production and standing stock estimates of the most abundant species, may represent also potential interesting tools to use on beach management and assessment practices. Nevertheless, functional traits remain largely unexplored by beach ecologists. Macrofaunal communities of exposed beaches, for instance, are largely dependent upon the stranded wrack brought ashore by tides, both as food and as habitat. Also, beach grooming is nowadays a common practice in beaches heavily explored by tourism all over the world, and is recognised as one important stressor with significant ecological consequences on the macrofauna (as reviewed in Defeo et al., 2009). These aspects reinforce the need for the development of ecological indicators for sandy beach assessment and management based on community functional attributes, a promising subject for further investigation.

4. Potential Key Species in Worldwide Sandy Beaches

Regarding management practices, the supralittoral and the intertidal zones can be considered priority areas for intervention on the beach environment. In urbanized coasts these areas are directly subjected to the detrimental impacts of several human activities, like for instance the construction of several types of urban and touristic facilities and mechanical and human trampling. Also, the environmental impacts resulting from on-going climatic changes, such as the rise of the sea level and the increase in storm occurrence and intensity, which promote for instance coastal squeezing, are important additional stressors that directly disturb these two zones. On the other hand, the largest proportion of the worldwide sandy shores is composed of open oceanic sandy beaches (McLachlan and Brown, 2006), although several physical aspects, like for instance the degree of exposure to wave action, may oscillate according to the local conditions and alter both the beach type and the species composition of their macrofaunal benthonic communities. These two assumptions will therefore guide us throughout this endeavour to identify the most relevant macrofaunal groups holding potential key species in worldwide sandy beaches.

Crustaceans are the most abundant macrofaunal organisms in exposed beaches and dominate the upper tidal levels (McLachlan and Brown, 2006). Taxonomic groups holding potential key species, according to our previous assumptions, are the amphipods, isopods and the brachyurans (table I). Within each one of these groups, several possibilities arise.

Table I. Summary of the potential key elements in the macrofauna of worldwide open oceanic sandy beaches, considering only supralittoral and intertidal inhabitants

Macrofaunal group	Habitat	Distribution	Genus	Example of known key or indicator species
Talitrid Amphipods	Supralittoral Intertidal	Temperate regions	*Talitrus*	*Talitrus saltator* (Gonçalves et al., 2009)
			Talorchestia	*Talorchestia brito* (Gonçalves et al., 2009)
			Orchestia	
			Orchestoidea	
Tylid Isopods	Supralittoral	Circumglobal	*Tylos*	*Tylos europaeus* (Gonçalves et al., 2009)
Cirolanid Isopods	Intertidal Subtidal	Widespread	*Eurydice*	
		Ubiquitous	*Excirolana*	*Excirolana braziliensis* (Veloso et al., 2011)
Ocypodid brachyuran crabs	Dunes Supralittoral Intertidal	Tropic and subtropic regions	*Ocypode*	*Ocypode quadrata* (Neves and Bemvenuti, 2006) *Ocypode cordimana* (Barros, 2001)
			Dotilla	
Polychaetes	Intertidal	Circumglobal	*Nephtys*	
			Scolelepis	

Talitrid amphipods, commonly known as sand hoppers, are characteristic of the supralittoral zones in temperate latitudes, although some species occupy also the intertidal. This family includes several genus that may be potential key elements, such as *Talitrus*, *Talorchestia*, *Orchestia* and *Orchestoidea*. In fact, *Talitrus saltator* and *Talorchestia brito* were recognized as key species in the western coast of Portugal by Gonçalves et al. (2009) and are believed to have similar roles in other exposed sandy shores of Europe and even North Africa, where they are known to be often abundant. Also, *Talitrus saltator* has been widely studied, especially on Atlantic and Mediterranean shores, and proposed as an ecological indicator species, namely for disturbances derived from human activities (see for instance Ugolini et al., 2008; Gonçalves et al., 2013; Bessa et al., 2013).

Regarding the isopods, Tylids and Cirolanids are the families with higher potential. The genus *Tylos* is noteworthy, since it has a circumglobal distribution from the tropics to cold-temperate latitudes (McLachlan and Brown, 2006), and holds several species that often constitute abundant populations in macrofaunal assemblages. *Tylos europaeus*, for instance, has been recognized as a key species in the western coast of Portugal and, like *T. saltator* and *T. brito*, may have similar roles in other Atlantic and Mediterranean beaches where it is often abundant (Gonçalves et al., 2009). Also, this species has recently been considered an ecological indicator tool in monitoring and assessment of the ecological quality condition of sandy beaches (Gonçalves et al., 2013).

Contrarily to Tylids which are typical of the upper-zones of the beach, Cirolanids are characteristic of the intertidal zone. In this family, the widespread *Eurydice* and the apparently ubiquitous *Excirolana* stand out as potential key elements in several worldwide coasts. Although for *Eurydice* very few attempts have been done to study the responses of these animals to human disturbances (but see Van Tomme et al., 2013) and its bioecology is poorly known, *Excirolana* is much better known and some specific responses to human disturbances have already been identified. *Excirolana braziliensis*, for instance, revealed to be a good species for monitoring anthropogenic impacts in Brazilian beaches due to its high resistance to environmental stress, persisting in highly urbanized areas dominated by bathers (Veloso et al., 2011).

Brachyurans, specifically Ocypodid crabs, are the most conspicuous members of the macrofauna in tropical and subtropical shores (McLachlan and Brown, 2006). The genus *Ocypode*, commonly known as ghost crabs, is one of the most studied ones and has been widely used as a reliable indicator of human disturbances in sandy beaches in the last three decades (see for instance Steiner and Leatherman, 1981; Barros, 2001; Neves and Bemvenuti, 2006; Lucrezi et al., 2008). This has been the case of, for instance, *Ocypode quadrata* on Brazilian and USA beaches and *Ocypode cordimana* on Indian and Pacific coasts. The genus *Dotilla* is another group that includes potential key species, since some species can be dominant in macrofaunal communities (see for instance Allen, 2010). However, studies analysing the responses of these crabs to anthropogenic stressors appear not to be available yet.

In exposed sandy beaches with high supplies of wrack deposits, Insects may be important macrofaunal residents on the supralittoral zone. With a few exceptions (e.g., Dugan et al., 2003; Gonçalves et al., 2009), little relevance has been given to these animals in sandy beach ecology studies. In our opinion, Insects, and namely Coleopterans, may also hold potential key species in beaches presenting these very specific environmental conditions, although their study is largely neglected. Further efforts should be done to improve our knowledge on insects bioecology, their role on macrofaunal assemblages, and to understand how they react to environmental disturbances, namely from human origin.

Although less common as dominants in macrofaunal communities of exposed beaches, molluscs and polychaetes also present some groups that may potentially hold key species in very specific conditions (table I). The bivalves *Donax*, for instance, are dominant on the intertidal and subtidal zones of beaches experiencing medium to strong water currents, almost all over the world (McLachlan and Brown, 2006). However, since *Donax* species are often commercially exploited and/or recreationally harvested, these populations may suffer large temporal variations in abundance and distribution (with consequences in other bioecological features) and therefore make it very difficult to use them as tools for assessment and management practices. Regarding polychaetes, the genus *Nephtys* and *Scolelepis* appear as promising ones. They are extremely widespread, from the tropics to the temperate regions, and tend to occupy distinct areas of the intertidal zone. In fact, some studies have already highlighted the relevance of polychaetes as effective and reliable ecological indicators in sandy beach ecosystems (e.g., Bergamino et al., 2009; Omena et al., 2012).

Conclusion

Based on the previous information that we presented, four distinct tools, relying on macrofaunal key species, are potentially available for sandy beach ecologists and managers: (1) population bioecology; (2) population dynamics models; (3) community structural attributes, and (4) community functional attributes. Until now, population biological and/or ecological features and community structural attributes have been, by far, the prevalent tools on sandy beach assessment, monitoring and management practices, although assessment and management strategies for these ecosystems are still very poorly developed. However, population dynamics models are very powerful tools that allow the beach manager/ecologist to simulate a multitude of distinct environmental scenarios and forecast the population's responses to the stressor introduced in the simulation. Also, human activities like beach grooming are responsible for considerably reducing the food and habitat availability for several macrofaunal inhabitants of exposed sandy beaches, altering the functional attributes of the community. Therefore, it is our recommendation that more effort should be spent with the development of these two promising tools.

On open oceanic sandy beaches, macrofaunal key species appear predominantly among crustaceans. Sand hoppers on temperate regions, and ghost crabs on the tropics and subtropics, are the groups that traditionally have been more studied and more often proposed as sensitive and reliable ecological indicators. However, as pointed out in the previous section, some isopods and a few other ocypodid crabs, which are not very well studied yet, may also hold potential key species. In this regard, we recommend that sandy beach ecologists should devote their first efforts on the study of macrofaunal communities, in an attempt to unveil if a key species is present. Only afterwards, should the research efforts be focused on the bioecology of the key species following, whenever possible, the steps indicated in section 3.1. However, it is our knowledge that this is not always possible, especially in highly urbanized coasts were undisturbed sandy beaches may sometimes be very difficult to find.

As a whole, since several manmade disturbances, and even natural disturbances may cause variations in key species bioecology, with negative impacts perpetuated to the higher ecological levels, monitoring, assessment and management actions focused on this indicator species hold large potential in the fast changing scenario of nowadays, and should be encouraged. As we said in the beginning of this commentary, there is still plenty to do regarding beach assessment and management practices, and even more on beach conservation, although we are nowadays advancing on beach research at a remarkably striking rhythm.

References

Allen, C. J. (2010). Ecology of the intertidal crab *Dotilla intermedia* from tsunami-impacted beaches in Thailand. University of Southampton, School of Ocean and Earth Science, Doctoral Thesis.

Anastácio, P. M., Gonçalves, S. C., Pardal, M. A., and Marques, J. C. (2003). A model for amphipod (*Talitrus saltator*) population dynamics. *Estuarine, Coastal and Shelf Science,* 58S, 149-157.

Barros, F. (2001). Ghost crabs as a tool for rapid assessment of human impacts on exposed sandy beaches. *Biological Conservation*, 97, 399-404.

Bergamino, L., Muniz, P., and Defeo, O. (2009). Effects of a freshwater. *Ecological Indicators*, 9, Issue 3, 584-587.

Bessa, F., Cunha, D., Gonçalves, S. C., and Marques, J. C. (2013). Sandy beach macrofaunal assemblages as indicators of anthropogenic impacts on coastal dunes. *Ecological Indicators*, 30, 196–204.

Caffyn, A., Prosser, B., and Jobbins, G. (2002). A Framework for the analysis of socio-economic impacts on beach environments. In: F. Scapini (Ed.) Baseline research for the integrated sustainable management of Mediterranean sensitive coastal ecosystems. A manual for coastal managers, scientists and all those studying coastal processes and management in the Mediterranean (37-50). Florence, Istituto Agronomico per l'Oltremare.

Colombini, I., Chelazzi, L. (2003). Influence of marine allochthonous input on sandy beach communities. *Oceanography and Marine Biology: an Annual Review*, 41, 115-159.

Davenport, J., and Davenport, J. L. (2006). The impact of tourism and personal leisure transport on coastal environments: a review. *Estuarine, Coastal and Shelf Science*, 67, 280-292.

Defeo, O., McLachlan, A., Schoeman, D. S., Schlacher, T. A., Dugan, J., Jones, A., Lastra, M., and Scapini, F. (2009). Threats to sandy beach ecosystems: a review. *Estuarine, Coastal and Shelf Science*, 81, 1–12.

Doney, S. C., Ruckelshaus, M., Duffy, J. E., Barry, J. P., Chan, F., English, C. A., Galindo, H. M., Grebmeier, J. M., Hollowed, A. B., Knowlton, N., Polovina, J., Rabalais, N. N., Sydeman, W. J., and Talley, L. D. (2012). Climate change impacts on marine ecosystems. *Annual Review of Marine Science*, 4, 11-37.

Dugan, J. E., Hubbard, D. M., McCrary, M. D., and Pierson, M. O. (2003). The response of macrofauna communities and shorebirds to macrophyte wrack subsidies on exposed sandy beaches of southern California. *Estuarine, Coastal and Shelf Science,* 58S, 25-40.

Fanini, L., Marchetti, G. M., Scapini, F., and Defeo, O. (2009). Effects of beach nourishment and groynes building on population and community descriptors of mobile arthropodofauna. *Ecological Indicators*, 9, 167–178.

Gonçalves, S. C., Anastácio, P. M., Pardal, M. A., Cardoso, P. G., Ferreira, S. M., and Marques, J. C. (2009). Sandy beach macrofaunal communities on the western coast of Portugal – Is there a steady structure under similar exposed conditions? *Estuarine, Coastal and Shelf Science*, 81, 555-568.

Gonçalves, S. C., and Marques, J. C. (2011). The effects of season and wrack subsidy on the community functioning of exposed sandy beaches. *Estuarine, Coastal and Shelf Science*, 95, 165-177.

Gonçalves, S. C. (2012). Crustaceans in Atlantic and Mediterranean exposed sandy beaches. Does knowledge in key species bio-ecology have a role in assessing global changes? Saarbrücken, Lambert Academic Publishing.

Gonçalves, S. C., Anastácio, P. M., and Marques, J. C. (2013). Talitrid and Tylid crustaceans bioecology as a tool to monitor and assess sandy beaches' ecological quality condition. *Ecological Indicators*, 29, 549-557.

Gormsen, E. (1997). The impact of tourism on coastal areas. *Geojournal*, 42 (1), 39-54.

Hazlett, B. A. (1988). Behavioural plasticity as an adaptation to a variable environment. In: G. Chelazzi, and M. Vannini (Eds.), Behavioural adaptation to intertidal life (317-332). London, Plenum Press.

Jones, A. R., Murray, A., Lasiak, T. A., and Marsh, R. E. (2008). The effects of beach nourishment on the sandy-beach amphipod *Exoediceros fossor*: impact and recovery in Botany Bay, New South Wales, Australia. *Marine Ecology,* 29, 28–36.

Leewis, L., van Bodegom, P. M., Rozema, L., and Janssen, G. M. (2012). Does beach nourishment have long-term effects on intertidal macroinvertebrate species abundance? *Estuarine, Coastal and Shelf Science*, 113, 172–181.

Lercari, D., and Defeo, O. (2003). Variation of a sandy beach macrobenthic community along a human-induced environmental gradient. *Estuarine, Coastal and Shelf Science*, 58, 17-24.

Lucrezi, S., Schlacher, T. A., and Walker, S. J. (2008). Monitoring human impacts on sandy shore ecosystems: a test of ghost crabs (*Ocypode* spp.) as biological indicators on an urban beach. *Environmental Monitoring and Assessment*, 152, 413-424.

Marques, J. C., and Anastácio, P. M. (2002). Integration and interpretation of ecological data at population level. In: F. Scapini (Ed.) Baseline research for the integrated sustainable management of Mediterranean sensitive coastal ecosystems. A manual for coastal managers, scientists and all those studying coastal processes and management in the Mediterranean (155-164). Florence, Istituto Agronomico per l'Oltremare.

Mayer, P. M. and Galatowitsch, S. M. (2001). Assessing ecosystem integrity of restored prairie wetlands from species production-diversity relationships. *Hydrobiologia*, 443, 177-185.

Mayer, P. M., Megard, R. O. and Galatowitsch, S. M. (2004). Plankton respiration and biomass as functional indicators of recovery in restored prairie wetlands. *Ecological Indicators*, 4, 245-253.

McLachlan, A. and Brown, A. C. (2006). The ecology of sandy shores (second edition). California, Elsevier.

McLachlan, A., Defeo, O., Jaramillo, E. and Short, A. D. (2013). Sandy beach conservation and recreation: guidelines for optimising management strategies for multi-purpose use. *Ocean and Coastal Management*, 71, 256-268.

Neves, F. M., and Bemvenuti, C. E. (2006). The ghost crab *Ocypode quadrata* (Fabricius, 1787) as a potential indicator of anthropic impact along the Rio Grande do Sul coast, Brazil. *Biological Conservation*, 133, 431–435.

Nordstrom, K. F. (2008). Beach and dune restoration. New York, Cambridge University Press.

Omena, E. P., Lavrado, H. P., Paranhos, R., and Silva, T. A. (2012). Spatial distribution. *Marine Pollution Bulletin*, 64, 9, 1861-1873.

Scapini, F., and Morgan, E. (2002). Bioassays for estimation of beach stability and ecosystem quality. In: F. Scapini (Ed.), Baseline research for the integrated sustainable management of Mediterranean sensitive coastal ecosystems. A manual for coastal managers, scientists and all those studying coastal processes and management in the Mediterranean (120-122). Florence, Istituto Agronomico per l'Oltremare.

Schlacher, T. A., Richardson, D., and McLean, I. (2008). Impacts of off-road vehicles (ORVs) on macrobenthic assemblages on sandy beaches. *Environment and Management*, 41, 878-892.

Schlacher, T. A., Noriega, R., Jones, A., and Dye, T. (2012). The effects of beach nourishment on benthic invertebrates in eastern Australia: Impacts and variable recovery. *Science of the Total Environment,* 435-436, 411–417.

Steiner, A. J., and Leatherman, S. P. (1981). Recreational impacts on the distribution of ghost crabs *Ocypode quadrata* fab. *Biological Conservation*, 20, Issue 2, 111-122.

Ugolini, A., Ungherese, G., Somigli, S., Galanti, G., Baroni, D., Borghini, F., Cipriani, N., Nebbiai, M., Passaponti, M., and Focardi, S. (2008). The amphipod *Talitrus saltator* as a bioindicator of human. *Marine Environmental Research*, 65, Issue 4, 349-357.

Van Tomme, J., Vanden Eede, S., Speybroeck, J., Degraer, S., and Vincx, M. (2013). Macrofaunal sediment. *Marine Environmental Research*, 84, 10-16.

Veloso, V. G., Silva, E. S., Caetano, C. H. S., and Cardoso, R. S. (2006). Comparison between the macroinfauna of urbanized and protected beaches in Rio de Janeiro state (Brazil). *Biological Conservation*, 127, 510-515.

Veloso, V. G., Neves, G., and Capper, L. A. (2011). Sensitivity of a cirolanid isopod to humans. *Ecological Indicators*, 11, Issue 3, 782-788.

Węslawski, J. M., Stanek, A., Siewert, A., and Beer, N. (2000). The sandhopper (*Talitrus saltator*, Montagu 1808) on the Polish Baltic coast. Is it a victim of increased tourism? *Oceanological Studies*, vol XXIX, no. 1, 77-87.

In: Beaches
Editor: Margaret Cessa

ISBN: 978-1-63117-239-7

Chapter 3

Improving Coastal Knowledge Transfer between Researchers and Managers: A Two-Way Route

M. M. Carapuço[1,*], *R. Taborda*[1,†], *C. Andrade*[2,‡] *and M. C. Freitas*[2,§]

[1]Instituto Dom Luiz, University of Lisbon, Campo Grande, Lisbon, Portugal

[3]Centro de Geologia da Universidade de Lisboa, University of Lisbon, Campo Grande, Lisbon, Portugal

---

* Corresponding author: M. M. Carapuço. Instituto Dom Luiz, University of Lisbon, Campo Grande, 1749-016 Lisbon, Portugal, Phone: +351 217 500 376, Fax: +351 217 500 065, E-mail: mmcarapuco@fc.ul.pt.

† R. Taborda: Instituto Dom Luiz and Departamento de Geologia, Faculty of Sciences, University of Lisbon, Campo Grande, 1749-016 Lisbon, Portugal, Phone: +351 217 500 376, Fax: +351 217 500 065, E-mail: rtaborda@fc.ul.pt.

‡ C. Andrade: Centro de Geologia da Universidade de Lisboa and Departamento de Geologia, Faculty of Sciences, University of Lisbon, Campo Grande, 1749-016 Lisbon, Portugal, Phone: +351 217 500 376, Fax: +351 217 500 065, E-mail: candrade@fc.ul.pt.

§ M. C. Freitas: Centro de Geologia da Universidade de Lisboa and Departamento de Geologia, Faculty of Sciences, University of Lisbon, Campo Grande, 1749-016 Lisbon, Portugal, Phone: +351 217 500 376, Fax: +351 217 500 065, E-mail: cfreitas@fc.ul.pt.

ABSTRACT

It is well recognized that the scientific knowledge on coastal dynamics should contribute to coastal zone management. However, so far, generation of data on coastal processes have been conducted almost exclusively by the research community; the data sets and results are made available in ways that suit that community, but frequently preclude direct application in coastal management. Moreover, there is a plethora of site-specific information that is frequently detained by manager, which is often overlooked by the research community. This chapter assesses both managers and researchers perspective on this subject. Results show that the major gap in knowledge transfer is related with the channels used for data dissemination, which are not adequate to reach managers. This stresses the importance of fostering the existing mechanisms to improve coastal knowledge transfer such as the use of coastal indicators, the existence of linkage agents, the development of "qualitative into quantitative" data tools and the development of managers-oriented tools to support self-data acquisition. Additionally, further meetings and seminars involving researchers and managers - "get together" mechanics - are also suggested by both parties. The perception of coastal researchers and managers concerning gaps in knowledge transfer reflects an unexpected general agreement; this proximity can constitute a major opportunity in bridging these communities together and in the implementation of a two-way route communication approach, thus increasing and improving the role of managers in the coastal knowledge transfer processes.

INTRODUCTION

The coastal zone is a complex and dynamic environment; beaches, in particular, are one of the most mutable environment in the world. Shorelines' strong variability in both time and space conflicts with its intense use or static occupation. Thus, coastal scientific knowledge should be integrated in the management of the coastal zone. However, there are gaps in the transfer of knowledge between coastal researchers and managers which constitute an obstacle in the interaction of this two parties and, thus, in the incorporation of scientific outcomes in the process of decision-making.

Knowledge transfer involves the processes for capturing, collecting and sharing explicit and tacit knowledge, including skills and competence (CEC, 2007) and a multitude of reasons for an ineffective transfer of knowledge between researchers and end-users have been suggested in literature.

Anderson (1992) attributes the reasons for this gap mainly to researchers who often allow much more interest, time, and effort to the production of new knowledge than to the dissemination of their research results. This then turns into a major barrier to the diffusion of research results to managers, policy-makers and practitioners.

According to van Koningsveld et al. (2005), the problem lays in the fact that frequently the data available are too complex and difficult to apply to the coastal management process - research findings cannot, or cannot easily, be put to practical use.

For other authors, the reasons of this gap are attributed to practitioners. Hence, the resistance of practitioners to adopt new knowledge such as outside sources or innovative suggestions, the motivation to seek and use information and their often-limited competencies and skills are some of the main handicaps for the appropriation and application of research results identified in literature (e.g.,, Hemsley-Brown and Oplatka, 2005; Kirst, 2010).

Regardless the nature of the gap, the reasons that led to this difficulty in communication have been identified by the scientific community based on their perception and believe of the facts; researchers have inferred from their experience the gaps in communication associated with, and due to, coastal managers. And the question that arises is if the scientific community has really identified the main reasons for the gap in transferring knowledge between coastal researchers and managers.

This chapter presents researchers and managers points of view concerning the topic of knowledge transfer. The fact is, whether the weakness - and strengths - is either on the researchers or practitioners side, it's largely admitted that knowledge transfer should be further encouraged and promoted. This can only be achieved by listen to both parties (Becheikh et al., 2010) and, subsequently, through the establishment and adoption of the proper mechanisms to overcome the gap and thus creating the needed conditions for a successful coastal knowledge transfer between researcher and managers in a two-way dialogue.

THE NEED TO LISTEN

Knowledge transfer involves sharing, among others, data and information (CEC, 2007). In a procedure of sharing information, it is assumed that the person or entity that generate knowledge knows the type and formats of the information to share and with whom. In most of coastal interventions, managers need scientific knowledge that may be translated into concrete actions. However, scientists and coastal managers frequently have different understandings (occasionally divergent) about the concept of relevant information: scientists every so often focus on the state of the art, and coastal managers expect decoded data and information to support management decision-making.

According to van Koningsveld et al., (2005) the explicit identification of the information that is needed to deal effectively with a given coastal problem is necessary to prevent miscommunication and divergence of perceptions. Thus, it is vital for an efficient scientific knowledge transfer to previously recognize the management objectives and consequent needs of information of coastal managers. Additionally, it is also important to identify the obstacles in communications between coastal researchers and managers, and that may constitute additional barriers to knowledge transfer.

In this sense, a web survey was conducted between September and November 2013, through the SurveyMonkey® platform, in order to allow coastal researchers and coastal managers to express their opinion concerning to knowledge transfer between these two parties. Results, which excludes personal data to ensure participants anonymity, are available at a University of Lisbon website (http://disepla.fc.ul.pt/ktransfer/index.htm). Data management and processing was done using IBM® SPSS® Statistics and ESRI® ArcGIS applications.

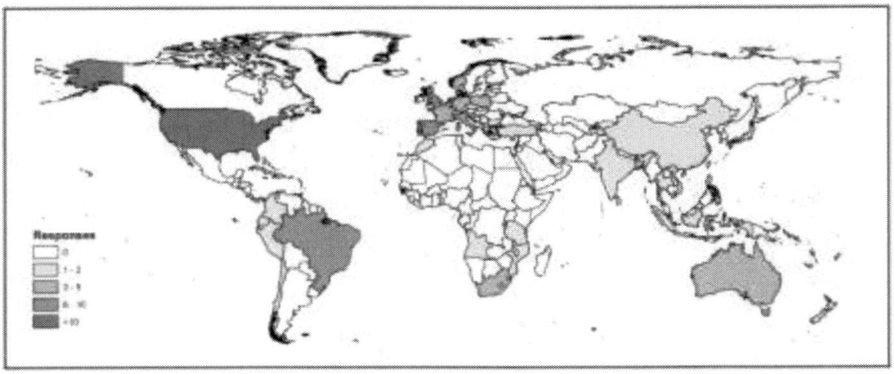

Figure 1. Geographic distribution of the researchers and managers sample used in this work.

Figure 1 presents geographical information concerning the target population based on the information of 174 enquiries, of which 107 were answered by coastal researchers and 67 were answered by coastal managers, who experience ranging from local to national authorities, non- to governmental organizations, tourism, fisheries and port authorities.

KNOWLEDGE GENERATION

Traditionally, knowledge generation is related to the research community. In this context, three types of research can be identified (van Koningsveld, 2003).

- Free research: the one conducted outside of the constraints of a practical problem and usually have no clearly identified end-users; the knowledge generate is focused in the state of the art.
- Applied research: seeks to solve practical problems; the knowledge generate is strictly end-user oriented and focuses on quick, efficient and economical solutions to practical problems.
- Driven research: when both practical and scientific interests drive the research in some form.

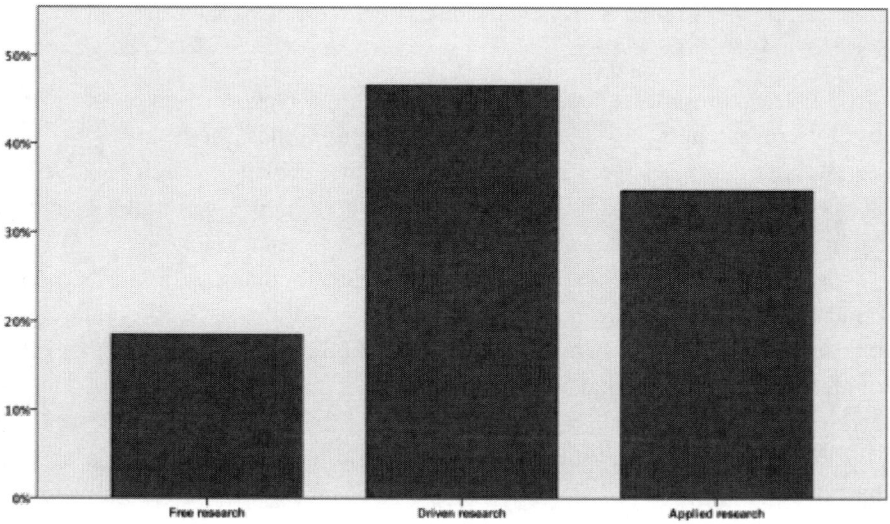

Figure 2. As a coastal researcher, how do you describe your research?

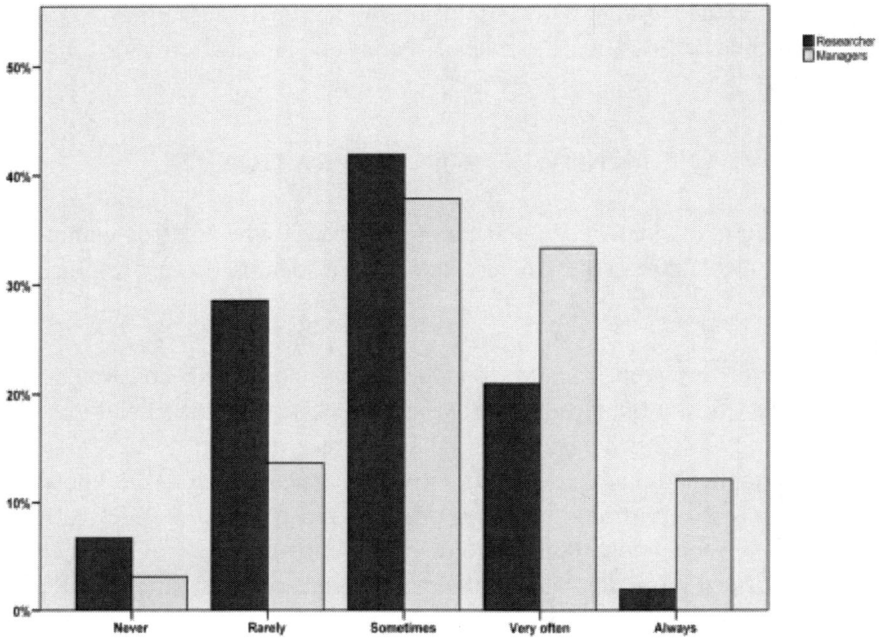

Figure 3. Do you work in cooperation with: *a) researchers:* coastal managers? *b) managers*: coastal researchers?

Figure 2 presents the results concerning to the type of research performed by the researchers. It is possible to conclude that the majority of the researchers that participate in the survey conduct driven research followed by applied research. Thus, it is possible to infer that the generated scientific knowledge is already incorporated in coastal zone management.

In realty, figure 3 (for which researchers that conduct free research did not contribute) shows that researchers and managers often work in collaboration although the manager's perception of this collaboration is slightly higher (40% of the managers refer that work in collaboration with research very often or always and researchers do not reach the 30%).

Figure 4 reveals that the incorporation of scientific knowledge is became part of the "business as usual"; around 40% of managers enquired always use scientific knowledge to support their activities. Still, there are opportunities for improving the integration of scientific knowledge in coastal management since 19% of managers use it sometimes and 8% use it rarely. The latter are managers that work at a local scale mainly in the public sector.

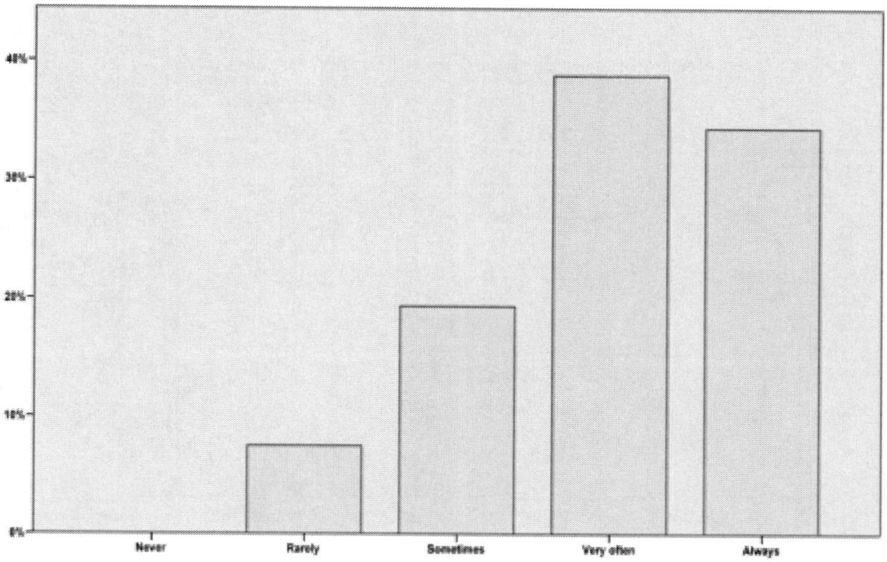

Figure 4. As a coastal manager, do you use scientific knowledge in your coastal management activities?

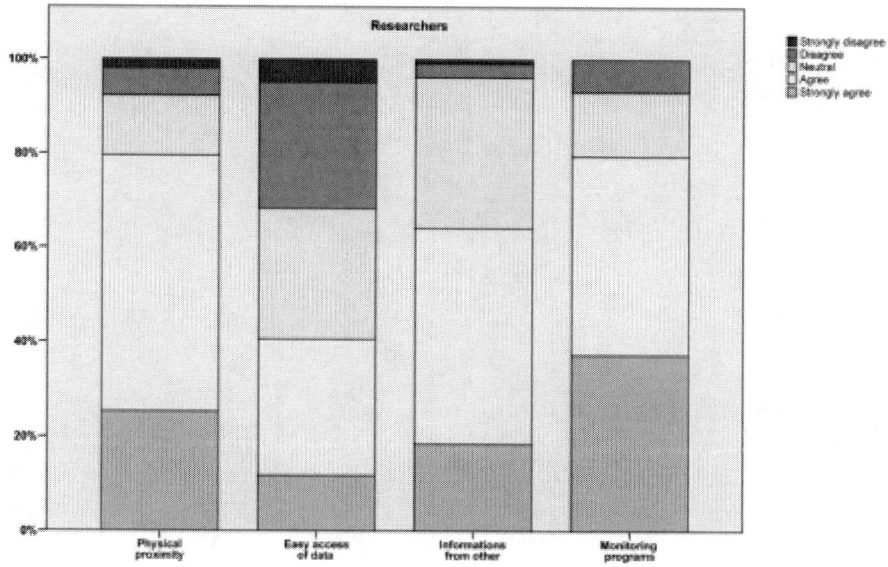

(a)

Figure 5. (Continued).

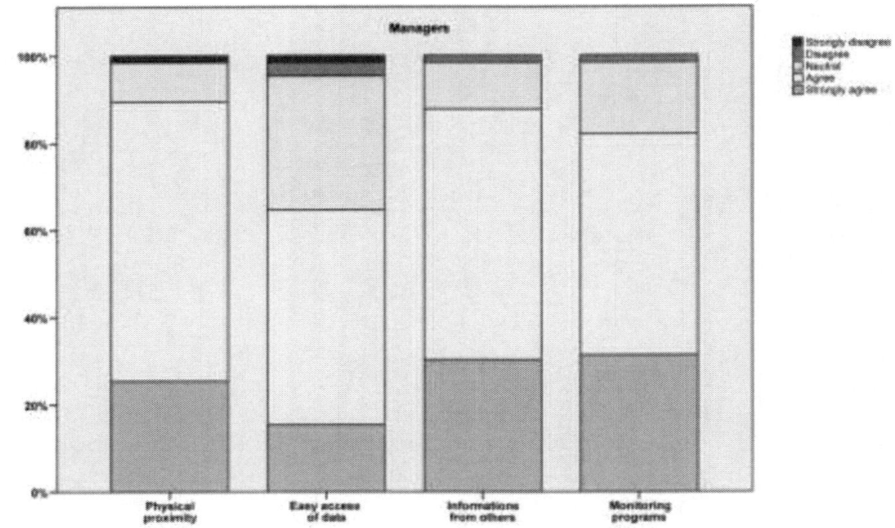

(b)

Figure 5. Choose the main reason(s) why you consider that: *a) researchers*: your research may benefit from the information detained by coastal managers. *b) managers*: your information may benefit coastal researchers activities.

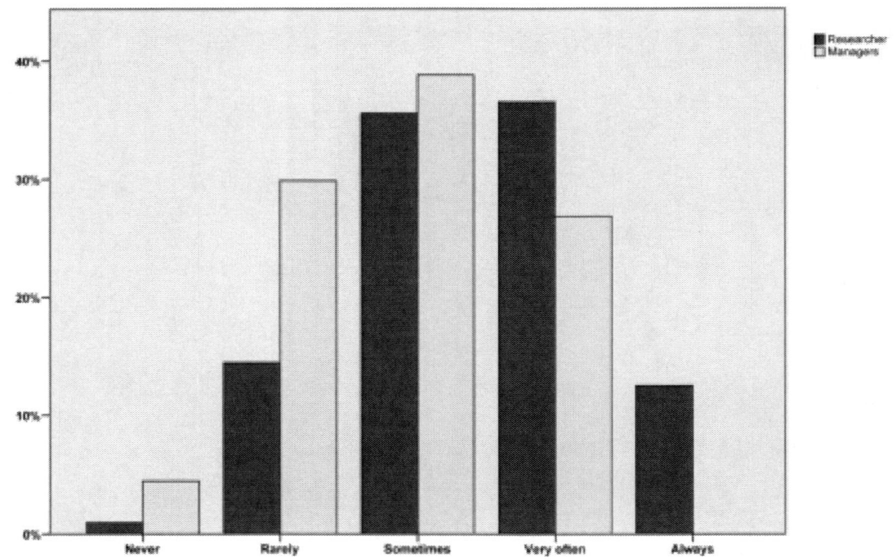

Figure 6. Coastal managers often have a lot of site-specific information. *Researchers*: Do you consider this type information in your research activities? *Managers*: How often does the research community ask you for this type of information?

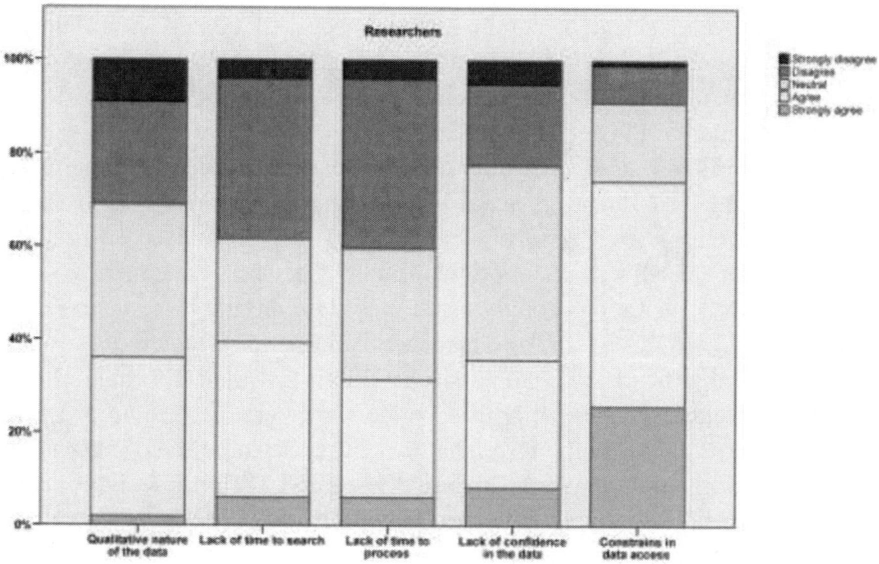

(a)

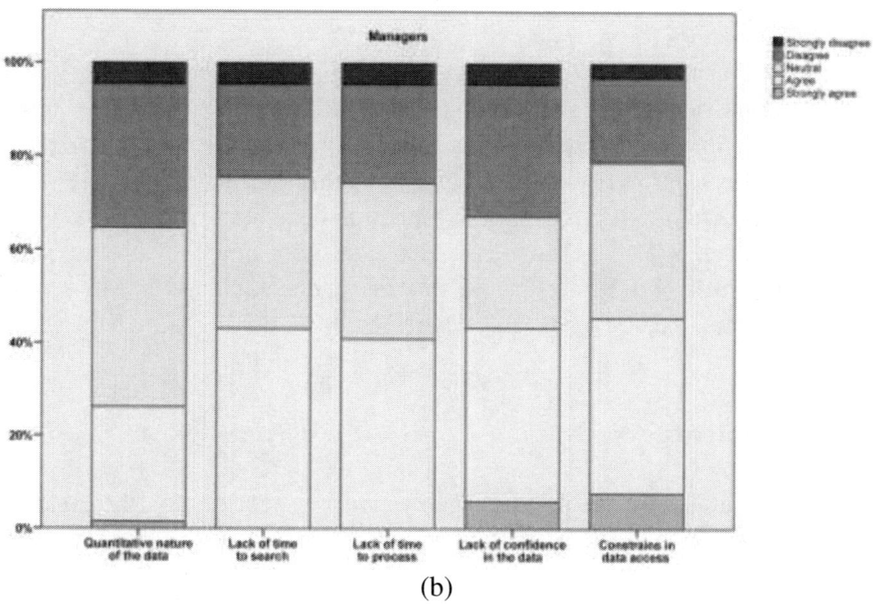

(b)

Figure 7. Choose the main reason(s) why you consider that: *a) researchers*: your research does not integrate more often site-specific information detained by coastal managers. *b) managers*: coastal researchers do not integrate more often site-specific information detained by coastal managers.

Traditionally research community is seen as primer contributor to knowledge generation; however, coastal managers contribution should not be overlooked. Actually, coastal managers detain a lot of site-specific data essential to support the decision making process. In fact, researchers recognize that manager's physical proximity to the coast, the existence monitoring programs, and since managers act as intermediates and receive information for others (e.g.,, beach user, fisheries' community) they are indeed an important source of site-specific information which might benefit researchers activities (more than 60% agree or strongly agree that these are the three main reasons why researchers may benefit from managers information) (Figure 5a).

On the other hand, managers considered also that their information might benefit coastal researchers activities for the same reasons (Figure 5b) (more than 80% agree or strongly agree that these are the three main reasons why managers information may benefit coastal researchers activities). In this sense, researchers often integrate site-specific information detained by coastal managers in their research activities (more than 50% use this type of data very often or always); still, managers considered that this information is not asked as often as it could, since more than 30% of managers claimed that researchers rarely or never ask for their data (Figure 6).

According to figure 7, what may justify that researcher do not take the maximum advantage of the site-specific information detain by coastal managers, is due to constrains in data access (Figure 7a); more than 70% of the researchers agree or strongly agree that this the major obstacle in the integration more often site-specific information detained by coastal managers. Coastal managers also acknowledge this constrain although also identified researchers lack of time to search, process and confidence in site-specific information detained by managers (Figure 7b).

Key Statements

- Traditionally knowledge generation it related to the research community.
- The majority of the researchers identifies themselves as driven or applied researchers; thus, it is possible to infer that mostly of the scientific knowledge generated aims to be incorporated in coastal zone management.

- Researchers acknowledge the importance of site-specific information detained by coastal managers; however, managers considered that this information is not asked as frequently as it could.
- Coastal managers can be considered knowledge generators as they detain and have facilitated access to site-specific information.
- Researchers do not take the maximum advantage of the site-specific information detained by coastal managers mainly due to constrains in data access.

KNOWLEDGE TRANSFER

Knowledge transfer can be described as a process of communication and translation (Liyanage et al., 2009). The standard communication process between two actors is usually described as a message transfer from an initiator to a receiver (Figure 8). In context of coastal knowledge, researchers are usually related to the initiator and mangers related to the receiver. Conceptually, this process is a one-way linear approach.

However, some authors refer that the communication process it is seldom in the way it occurs in practice and each step in the process often requires continuous iterations and ongoing collaborations among relevant parties (e.g.,, Becheikh et al., 2010; Sudsawad, 2007).

Figure 9a presents the effort/time put in the different "means of communication" by researchers to disseminate their knowledge on coastal processes. Research papers and technical reports are the one were researchers put more effort while non-technical outputs received less attention. The prefer means that managers use to access to scientific follows a similar trend (Figure 9b).

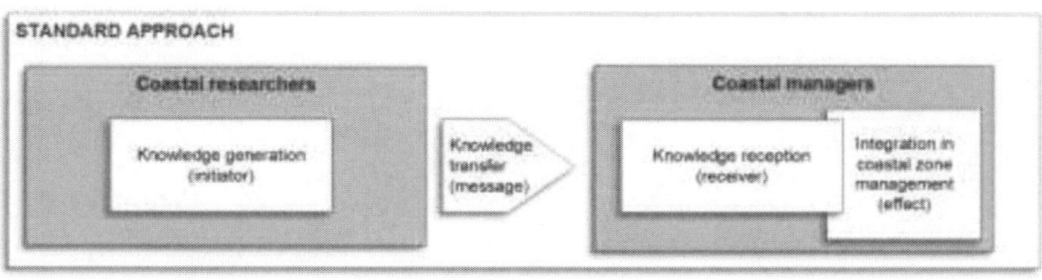

Adapted from Becheikh et al., 2010.

Figure 8. Knowledge transfer process standard approach.

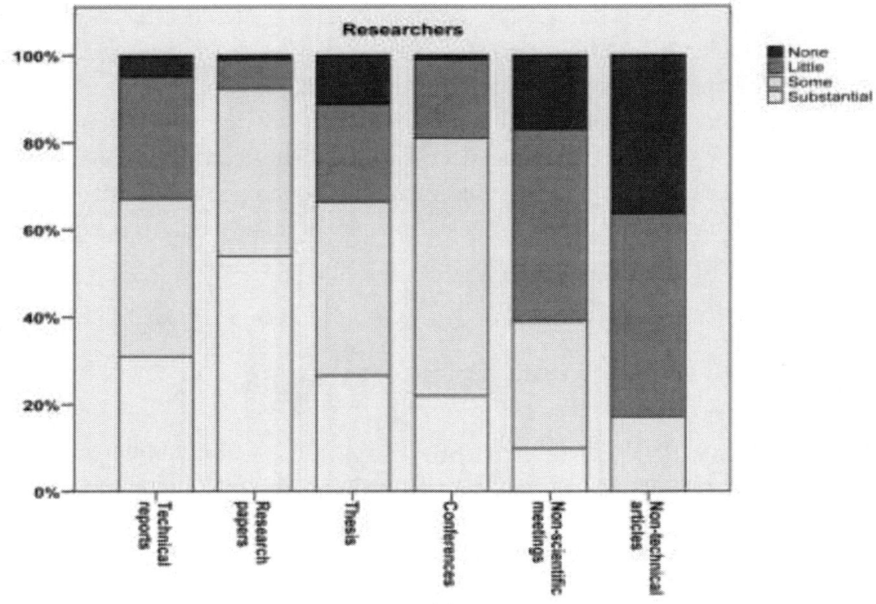

(a)

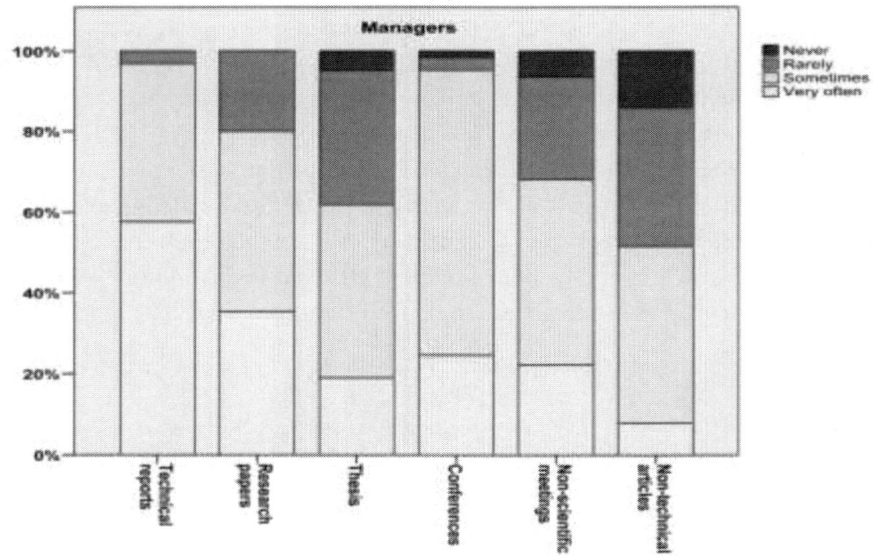

(b)

Figure 9. Data dissemination. a) *researchers*: How much effort/time do you spend with each of these means of scientific knowledge dissemination? *b) managers*: What are the means that you use to access scientific information?

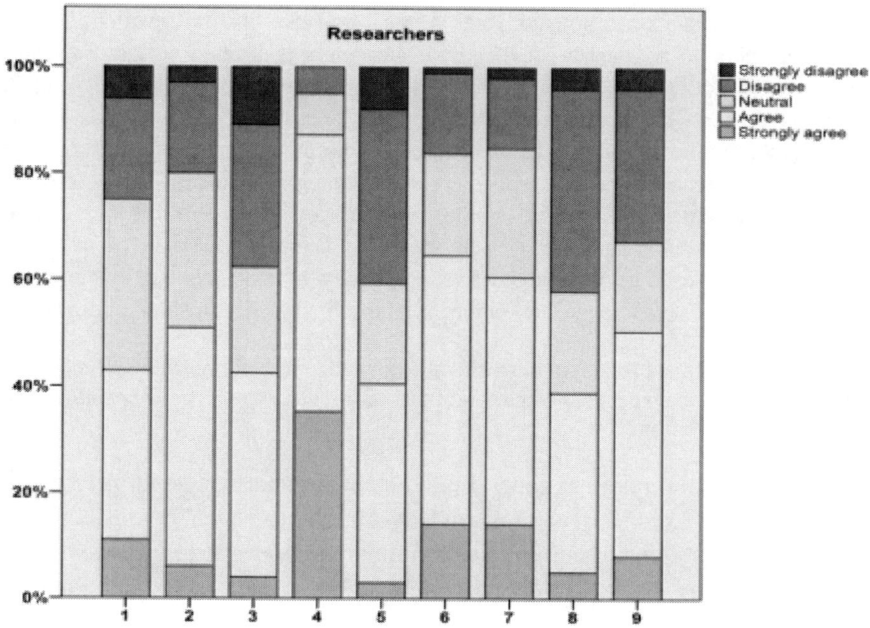

(a)

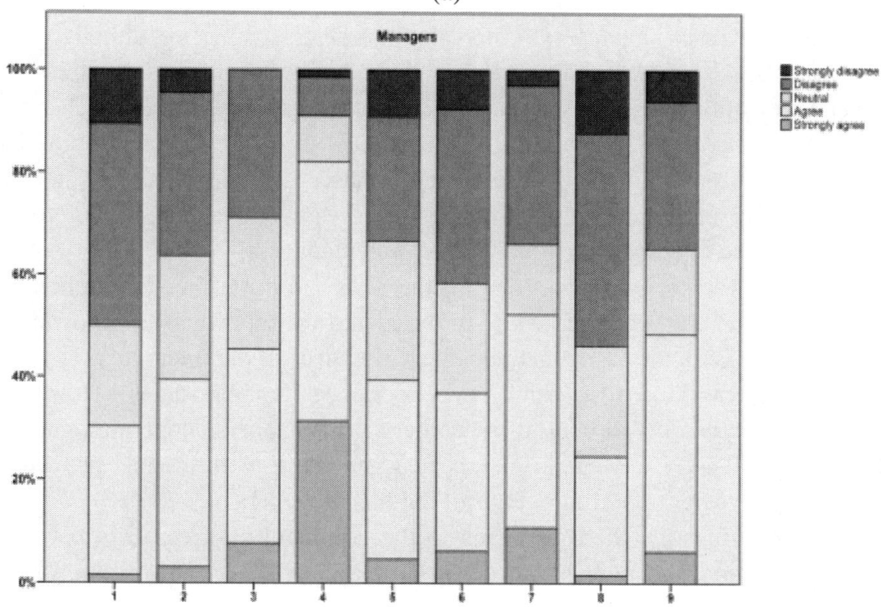

(b)

Figure 10. (Continued).

1. Coastal managers are not motivated to seek and use new information. 2. Coastal managers do not clearly identify their management needs. 3. Coastal researchers are not motivated to share information. 4. The channels used in data dissemination (e.g., scientific articles) are not adequate to reach managers. 5. The data sets and results are too complex to be understood and used by coastal managers. 6. There is a lack of managers' competencies to integrate scientific knowledge in management. 7. There is a lack of managers' time to integrate scientific knowledge in management activities. 8. There is a lack of researchers' competencies to generate and deliver useful data to managers. 9. There is a lack of researchers' competencies to traduce scientific knowledge in simple information.

Figure 10. Which are the main obstacles to knowledge transfer between coastal researchers and managers: *a) researchers*: based in your experience. *b) managers*: based in your opinion.

These results revels a general agreement in what concerns to researchers and managers perspective on knowledge transfer, which could suggest the absence of problems in communication amongst both parties. However, when directly asked, obstacles in communication are recognized. Figure 10 highlights that the major obstacle identified by both parties it is the channels used in data dissemination (e.g.,, scientific articles), which may not be adequate to reach managers. This suggests that managers, although use scientific articles and technical reports to access to scientific data, they are indeed constrained to use the available means. At the same time, since researchers also acknowledge the channels used in data dissemination as the major obstacle in knowledge transfer, it raises issues concerning to the "policy" of scientific data dissemination and researcher evaluation that may be contributing to "getting apart" researchers and managers.

Figure 10 also revels that among the same group (either researchers or managers), the answers concerning to the identification of gaps in knowledge transfer is quite diverse; this may be due differences in the interveners´ profile (e.g.,, different skills, responsibilities, experiences, competencies). However, the overall perception of coastal researchers and managers concerning gaps in knowledge transfer reflects an unexpected general agreement; this proximity constitutes a major opportunity in bridging them together.

Recognizing that there are gaps in the communication process between researchers and managers, that have different profiles (backgrounds and experiences), an effective knowledge transfer should be based on tools to decode information. Professional organizations have put forth that indicators constitute the most efficient way to transmit inherently complex information in a simplified and applicable form (e.g.,, UNESCO, 2003; UNESCO, 2006;

NOAA, 2010). Other authors identify linkage agents (intermediary actors between researchers and managers also designated as knowledge-brokers, boundary-spanners, gatekeepers, credible messenger or translators) as the proper mechanism to enhance and improve the knowledge transfer process between two communities (e.g.,, Huberman, 1987; Huberman, 1990; Lavis et al., 2003).

In this study, additionally to these mechanics, the development of tools that allow coastal researchers to transform end-users qualitative data into quantitative information, and the development of managers-oriented tools to support self-data acquisition, were also considered. Figure 11 reveals that researchers and managers considered coastal indicators an adequate tool to improve scientific knowledge transfer (in both cases, more than 80% agree or strongly agree that coastal indicators can improve coastal knowledge transfer).

However, there is a general agreement (in both cases, higher than 60%) that all the other tools can improve knowledge transfer. The major difference between researchers and managers concerns the development of managers-oriented tools to support self-data acquisition; managers are willing to increase independency in self-acquisition of data to support coastal management (ca. 90% agree or strongly agree with option) whereas researchers seem more reluctant (ca. 65% agree or strongly agree with this option).

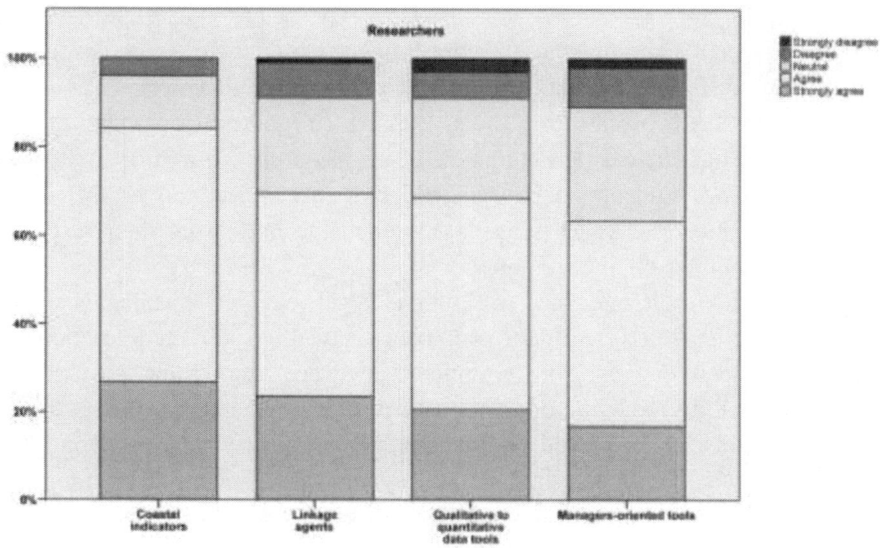

(a)

Figure 11. (Continued).

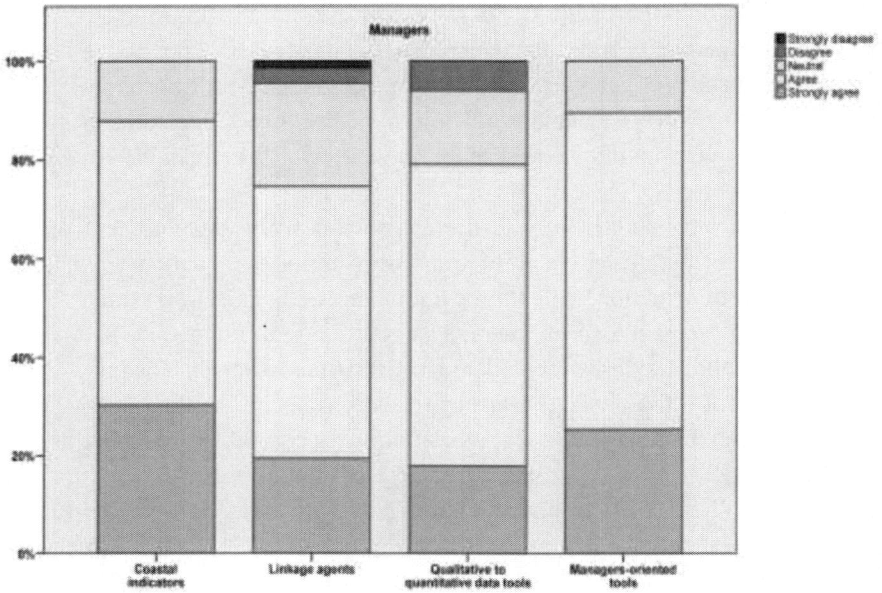

(b)

Figure 11. Which of the following could improve coastal knowledge transfer between researchers and managers? *a) researchers*: based in your experience. *b) managers*: based in your opinion.

Based on the commentaries concerning to this issue, two reasons can justify this difference: researchers invoked the risk of that option being used by managers as "black box" tools, lacking full acknowledgment of their limitations, thus increasing the chance of using outputs in improper ways; on the other hand, some managers claimed that this option can be difficult to implement since researches may be apprehensive in sharing their skills and losing part of their distinctive value.

Beside the tools presented in figure 11, "get together" mechanics such as regular meetings and seminars promoted as well as directed to, both the research and the managers communities, were also refer as adequate mechanisms to improve the interaction between the two parties.

Given this, an improved communication approach is proposed in figure 12. In this scheme, the communication process takes place in two directions: towards coastal managers but also towards coastal researchers. Since managers were identified as an important, and is some aspects unique, source of information they must also being involved in the process of knowledge generation and transfer.

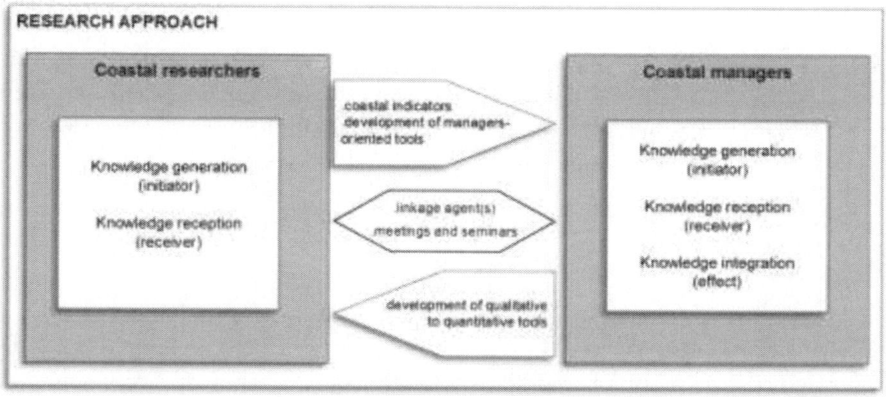

Figure 12. Knowledge transfer process research approach.

Key Statements

- The standard communication process between two actors usually consists in a message transfer from an initiator (researcher) towards a receiver (manager); this so called linear approach is a one-way route that does not take advantage of knowledge detained by managers.
- The communication process will benefit from adopting a two-way route approach.
- The major gap in knowledge transfer identified by both researchers and managers is the channels used for data dissemination, which are not adequate to reach managers.
- Coastal managers use technical reports and research papers to access scientific information in the absence of more friendly or targeted alternatives.
- The actual “policy” of scientific data dissemination does not foster knowledge transfer between researchers and managers.
- Amongst the same group (either researchers or managers), the obstacles identified in knowledge transfer are very dissimilar which shows that it is not possible to adopt or recommend a universal approach.
- The perception of coastal researchers and managers concerning gaps in knowledge transfer reflects an unexpected general agreement; this proximity constitutes a major opportunity in bridging them together.

- Translation tools and "get together" mechanisms should be improved; in this sense, the use of coastal indicators, linkage agents, meetings and seminars, "qualitative into quantitative" data tools and managers-oriented tools to support self-data acquisition, is recommend by both researchers and managers.

Conclusion

Assuring that coastal research is a source of application-oriented knowledge is an ambition shared by researchers and managers. Achieving this objective implies an effort to overcome existing gaps in the communication process. The results presented herein highlight the need to attend to the specificity of the actors and thus it is not possible to adopt or recommend a universal solution.

Although the standard communication process is usually liner and takes place in a one-way route, since coastal managers can also be relevant knowledge generators, knowledge transfer will strongly benefit from adopting a two-way route approach.

In the translation of scientific information into an applicable form by managers, the use of coastal indicators, the existence of linkage agents and "qualitative into quantitative" data tools were identified as adequate mechanisms. Managers-oriented tools to support self-data acquisition and further meetings and seminars involving researchers and managers - "get together" mechanisms - are also suggested by both researchers and managers.

The perception of coastal researchers and managers concerning the knowledge transfer process reflects a surprising general agreement between the answers; this proximity can constitute a major opportunity in bridging these communities.

Acknowledgments

The authors are grateful for the contributions of Tanya Silveira, Norb Psuty and Célia Ferreia in the elaboration of the enquires as well as to all the participants in the web survey. The first author is supported by a PhD grant SFRH/BD/82223/2011 funded by the Portuguese *Fundação para a Ciência e Tecnologia* – FCT. This work is a contribution of the projects Beach to

Canyon Head Sedimentary Processes (PTDC/MAR/114674/2009) funded by FCT and "Criação e implementação de um sistema de monitorização no litoral abrangido pela área de jurisdição da Administração da Região Hidrográfica do Tejo" funded by the National Environmental Agency (Agência Portuguesa do Ambiente, I.P.).

REFERENCES

Anderson, B. L. 1992. *Successful Curriculum Reforms: Sharing the Knowledge with Policymakers and Practitioners in Ways That Influence Practice*. Office of Educational Research and Improvement, Washington, DC. Office of Research.

Becheikh, N., Ziam, S., Idrissi, O., Castonguay, Y., Landry, R. 2010. How to improve knowledge transfer strategies and practices in education? Answers from a systematic literature review. *Research in Higher Education Journal,* 7: 1-21.

CEC (Commission of the European Communities) 2007. Improving knowledge transfer between research institutions and industry across Europe: embracing open innovation. *COM(2007) 182 final*. Brussels.

Hemsley-Brown, J. and Oplatka, I. 2005. Bridging the research-practice gap: barriers and facilitators to research use among school principals from England and Israel. *International Journal of Public Sector Management,* 18(5): 424 – 446.

Huberman, M. 1987. Steps toward an integrated model of research utilization. *Knowledge: Creation, Diffusion, Utilization,* 8: 586-611.

Huberman, M. 1990. Linkage between researchers and practitioners: A qualitative study. *American Educational Research Journal,* 27(2): 363-391.

Kirst, M. W. 2010. Bridging education research and education policymaking. *Oxford Review of Education* 26(3/4): 379-391.

Lavis, J., Robertson, D., Woodside, J., McLeod, C., Abelson, J. 2003. How can research organizations more effectively transfer research knowledge to decision makers. *Milbank Quarterly*, 81 (2): 221–248.

Liyanage, C., Elhag, T., Ballal, T., Li, Q. 2009. Knowledge communication and translation – a knowledge transfer model. *Journal of Knowledge Management,* 13(3):118 – 131.

NOAA (National Oceanic and Atmospheric Administration), 2010. *Coastal Zone Management Act - Performance Measurement System: Contextual Indicators Manual*, 38 p.

Sudsawad, P. 2007. *Knowledge translation: Introduction to models, strategies, and measures.* Austin, TX: Southwest Educational Development Laboratory, National Center for the Dissemination of Disability Research.

UNESCO (United Nations Educational, Scientific and Cultural Organization), 2006. A handbook for measuring the progress and outcomes of integrated coastal and ocean management. *IOC Manuals and Guides*, 46, ICAM Dossier 2, 224 p.

UNESCO (United Nations Educational, Scientific and Cultural Organization), 2003. A reference guide on the use of indicators for integrated coastal management. *IOC Manuals and Guides*, 45, ICAM Dossier 1, 127 p.

Van Koningsveld, M., 2003. Matching Specialist Knowledge with End User Needs. *PhD Thesis*, University of Twente, Enschede, The Netherlands.

Van Koningsveld, M., Davidson, M. and Huntley, D., 2005. Matching Science with Coastal Management Needs: The Search for Appropriate Coastal State Indicators. *Journal of Coastal Research*, 213: 399–411.

In: Beaches
Editor: Margaret Cessa
ISBN: 978-1-63117-239-7

Chapter 4

The Influence of Environmental Aspects in the Variation of Natural Radioactivity Levels in Selected Brazilian Beach Sand Samples

Lívia F. Barros[1,*] and Brigitte R. S. Pecequilo[2,†]
Instituto de Pesquisas Energéticas e Nucleares
(IPEN-CNEN/SP), São Paulo, SP, Brazil

Abstract

At beaches, environmental implications added to shoreline erosion and artificial interventions on native structure may alter the concentration of the natural radionuclides in certain regions of the coast. In this work, eleven locations at Camburi beach, known to be a naturally high background region, were studied in 2011. The influence of environmental aspects, as geology, average temperatures, variations of the height of the tides and the rainfall levels were related with the variation of ^{226}Ra, ^{232}Th and ^{40}K activities concentrations, during the year 2011 and annual effective dose were evaluated from the ^{226}Ra, ^{232}Th and ^{40}K sands activities concentrations.

* Lívia F. Barros: Instituto de Pesquisas Energéticas e Nucleares (IPEN-CNEN/SP), Av. Professor Lineu Prestes 2242, Cidade Universitária, 05508-000 São Paulo, SP, Brazil. E-mail: lfbarros@ipen.br.
† Brigitte R. S. Pecequilo: E-mail: brigitte@ipen.br.

Sand samples were monthly collected during 2011, dried, sealed in standard 100 mL HPDE polyethylene flasks and measured by high resolution gamma spectrometry after a 4 weeks in-growth period. The results, considering samples gamma-rays self-attenuation, show activities concentrations in the range from 4 Bq kg^{-1} to 1380 Bq kg^{-1} for ^{226}Ra, 9 Bq kg^{-1} to 7453 Bq kg^{-1} for ^{232}Th, and 6 Bq kg^{-1} to 504 Bq kg^{-1} for ^{40}K. The absorbed dose rate (D) in outdoor air values laid from 9 nGy.h^{-1} to 5160 nGy.h^{-1} .The annual effective dose values for the sand samples considering a daily exposure of 24 hours all year round laid from 0.05 mSv.y^{-1} to 31.64 mSv.y^{-1}. The annual effective dose values for the sand samples considering daily exposure of one hour all year round laid from 0.002 mSv.y^{-1} to 1.318 mSv.y^{-1}.The annual effective dose (E) for the beach collection points for a resident of Vitoria city, with one hour daily exposure all year round by collection point laid from 0.003 mSv.y^{-1} to 0.232 mSv.y^{-1}. The annual effective dose (*E*) for the collection points from Camburi beach for a tourist with 8 hours daily exposure on january, july or both, laid from 0.004 mSv.y^{-1} to 0.446 mSv.y^{-1}. For a resident of Vitoria city and for a tourist, in the conditions of exposure of this work, all values of *E* are lower than the limit of 1 mSv y^{-1}, recommended for ICRP 60 for the general public. In some months of the year 2011, low average temperatures, low rainfall levels and low variations of the height of the tides can be related with high ^{226}Ra, ^{232}Th and ^{40}K activities concentrations. In April and in October, high rainfall levels and high variations of the height of the tides can be related with low ^{226}Ra, ^{232}Th and ^{40}K activities concentrations.

1. Introduction

Natural radioactivity is always present in the earth and it exists in different levels in the entire environment. Naturally occurring radionuclides of terrestrial origin (also called primordial radionuclides) are present in various geological formations in beach sand, rocks and soils. Only those radionuclides with half-lives comparable to the age of the earth, and their decay products, as ^{40}K and radionuclides from the ^{238}U and ^{232}Th series, exist in significant quantities to contribute to the dose to mankind [1].

Throughout the history, gamma ray exposure from ^{40}K and radionuclides from the ^{238}U and ^{232}Th series occurs naturally in the nature [1]. According to UNSCEAR [1], about 87% of the radiation dose received by mankind is due to natural radiation sources.

The main contribution to the external exposure comes from gamma emitters radionuclides in soils, especially the ^{40}K and the ^{238}U and ^{232}Th series.

These radionuclides are also present in the body (ingested or inhaled) and they irradiate various organs with alpha and beta particles and gamma radiation [1]. Previous studies showed high activities concentrations of ^{226}Ra (from ^{238}U serie), ^{232}Th and ^{40}K in beach sands located in regions with levels of exposure due to natural sources of radiation above those considered "normal background". It is the case of beach sands of Egypt [2, 3], China [4], Turkey [5], India [6], Bangladesh [7], Brazil [8, 9] and also in other countries like Australia, Austria, France, Iran, South Africa , Thailand and Sri Lanka [10].

In Brazil, the levels of radioactivity of high background are mainly due to rare earth elements found, occurring in monazite and in other heavy minerals such as ilmenite, rutile and zircon, found in beach sands [10].

The ^{226}Ra, ^{232}Th and ^{40}K are present in significant amounts in monazite, ilmenite, rutile and zircon, depending on the local geology of the region of interest.

Among the largest anomalies in the concentration of thorium in the soil in Brazil are deposits of monazite sands from Guarapari in the state of Espírito Santo [11]. The Areia Preta beach sands in Guarapari been studied in 2003, as the seasonal variation of their high levels of radioactivity [12]. A typical brazilian monazite contains 6% thorium oxide and 0.3% uranium oxide [13].

Recent researches [14-17] show that the Camburi beach sands contain monazite, ilmenite, rutile and zircon, rich in ^{226}Ra, ^{232}Th and ^{40}K and it present high levels of NORM (Normally Occurring Radioactive Materials).

The knowledge about these activities concentrations and the distribution of these natural radionuclides in environmental sites plays an important role in radiation protection [1].

Some environmental aspects can change the natural structure of a beach throughout a year. The dynamics in sand transport involves the winds, the waves, coastal currents and tides, which act continuously on the materials found on the beach, eroding, transporting and depositing sediments. The concentration of some minerals in the sands, the density, diameter and shape of the sand grains and others many factors related to dynamic beach may alter the concentration of these natural radionuclides in certain regions of the coast [14]. Higher values of activity concentration of ^{226}Ra may be related to a higher concentration of ilmenite, while higher values of activity concentration of ^{232}Th, a higher concentration of monazite [9].

The activities concentrations of ^{226}Ra, ^{232}Th and ^{40}K at a specific location on the beach are subject to considerable fluctuations in a time, in particular due the different concentrations of monazite, ilmenite, rutile and zircon on the sands throughout a year.

The main objective of this study is to investigate the influence of some environmental aspects in the variation of natural radioactivity levels of ^{226}Ra, ^{232}Th and ^{40}K in locations at Camburi beach. This beach presents particular characteristics due to natural and artificial processes occurred during the process of urbanization in the region of Vitória.

2. Environmental Implications of Camburi Beach

Recent researches [14-17] show that the Camburi beach sands contain monazite, ilmenite, rutile and zircon, rich in ^{226}Ra, ^{232}Th and ^{40}K. Camburi beach is located in Vitória, capital of Espírito Santo State, Brazil.

On the beach, a pattern of dark spots occurs due to the presence of iron ore in sands (Figure 1).

The natural processes of deposition of these minerals include the influence by wind, by tidal and by the actions of waves and currents, in a dynamic which depends on the weather conditions beach.

The current morphodynamic of Camburi beach is directly related to environmental implications, as shoreline erosion and artificial interventions realized on its native structure [18].

Erosion resulting from the occupation process had generated several interventions until the present day, as the construction of piers perpendicular and artificial nourishment.

Photo: Personal archive.

Figure 1. Camburi beach. A speckled pattern in the sand is more visible in the central region of the beach and next to sea line.

There were already two processes of artificial nourishment, the first in 1980, which was not satisfactory and another in 1999, that obtained good results. In the latter, there was a withdrawal of 1 million cubic meters of sand from the submerged area adjacent to the beach, making it more reflective and stable [19].

Other interventions were the construction of Pier 1 and Pier 2, Pier of Iemanjá and the construction of the Tubarão Seaport with constant dredging - excavation in the access canals and berthing areas of the seaport to maintain depth- until the present day (Figure 2).

Temperature, height of the tides and the rainfall levels are climatic factors with seasonal behavior. Climate changes alter the dynamic related to the transportation of monazite, ilmenite, rutile and zircon along the beach. The concentration of these minerals to a single point of the beach can change throughout the year [14].

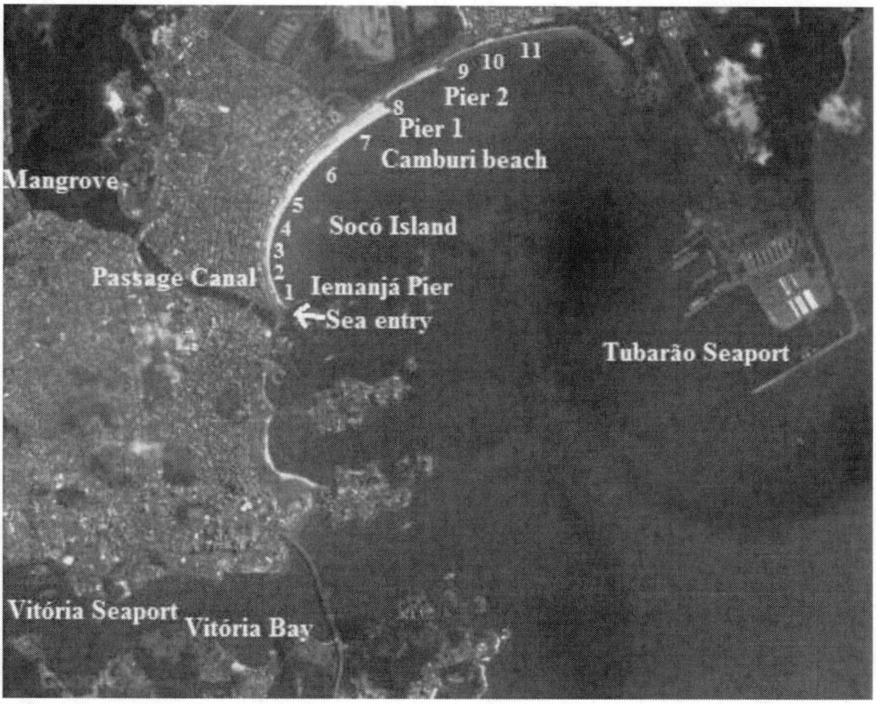

Figure 2. Camburi beach area. The collection points (1-11), Pier 1, Pier 2, Iemanjá Pier, Tubarão Seaport, Vitória Seaport and Vitória Bay are shown. The sea communicates with the mangrove through the Passage Canal [20].

3. Materials and Methodology

3.1. Study Area

Camburi beach is located in Vitória, capital of Espírito Santo State, Brazil, as shown in Figure 3. Camburi is the only beach located in the mainland city, with an extension of 6 km, with people attending the whole year. The weather is warm almost the entire year, so Camburi beach is constantly frequented both by locals and tourists.

3.2. Samples Collection and Preparation

For the study, eleven sampling points at Camburi beach were selected, between latitudes 20°15'58.29"S and 20°17'32.89"S and longitudes 40°15' 54.98"W and 40°17'20.85"W, covering the 6 km shore extension (Figure 4). The sand samples were collected almost at the beach surface, within a depth of about 2 cm (simulating a person sitting on the beach sand) and monthly, every 15th of the month, from January to December 2011. After manually removing macro-impurities (like shells, pebbles, food leftovers), the sand samples were dried on a stove and sealed in 100-mL HDPE flat-bottom cylindrical flasks with 52.5 mm plan screw cap and bubble spigot.

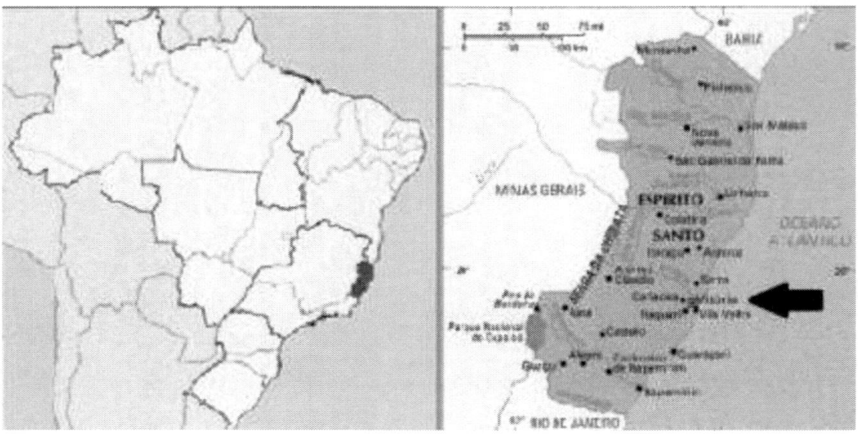

a b

Figure 3. (a) Brazil map (in black Espírito Santo State) [21]; (b) Espírito Santo State map (Vitória city indicated by the black arrow) [22].

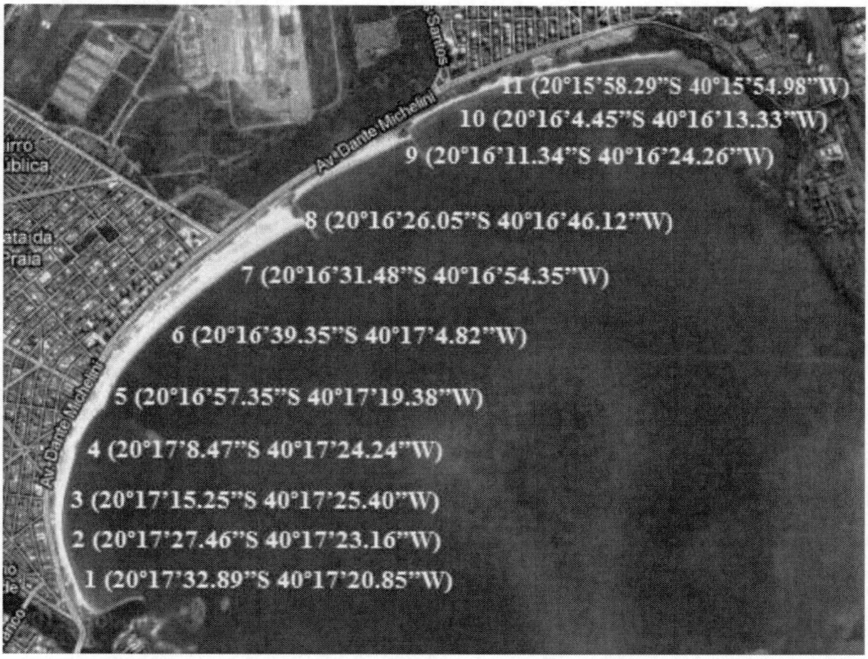

Figure 4. Camburi beach with Google geographical coordinates of the collection points [20].

3.3. Measurements

3.3.1. Activities Concentrations of ^{226}Ra, ^{232}Th and ^{40}K

The sand samples were measured by high resolution gamma spectrometry, after a 4 weeks in-growth period, in order to allow secular equilibrium in the ^{238}U and ^{232}Th series [23].

The detection system is a 15% HPGe ORTEC EG and G detector with conventional electronics and a 919 ORTEC EG and G Spectrum Master 4k multichannel analyzer. The effective resolution for the ^{60}Co 1332 keV energy is 2.8 keV.

The background radiation was obtained with ultrapure water, in the same sample geometry. The efficiency calibration curve was performed with a multinuclide standard aqueous solution in the same geometry as all measured samples.

All spectra were analyzed with the InterWinner 6.0 software [24], considering samples gamma-rays self-attenuation.

The activity concentration of a single transition was calculated as equation (1) [23]:

$$A_i(E_{\gamma i}) = \frac{C}{\varepsilon_{\gamma i}.\mathrm{I}_{\gamma i}.t_u.m} \times f(E_{\gamma i}) \tag{1}$$

where:

$A(E_{\gamma i})$ = activity of the considered gamma transition of the isotope in the sample
(Bq.kg^{-1});
C = net area for the gamma transition with energy ($E_{\gamma i}$) emitted by isotope;
$\varepsilon_{\gamma i}$ = detector efficiency for the considered gamma transition;
$I_{\gamma i}$ = probability of emission of the gamma transition with energy ($E_{\gamma i}$);
t_u= counting live time (s);
m = sample mass (kg);
$f(E_{\gamma i})$ = self-attenuation factor for the considered gamma transition.

The activity of ^{40}K was calculated through its single gamma transition of 1460.83 keV. Considering the radioactive equilibrium in the ^{238}U and ^{232}Th series, the activity of ^{226}Ra was determined by the weighted mean of the ^{214}Pb (295.21 keV and 351.92 keV) and ^{214}Bi (609.32 keV) gamma transitions and the activity of ^{232}Th by the weighted mean of the ^{212}Pb (238.63keV and 300.09 keV), ^{212}Bi (727.33 keV) and ^{228}Ac (911.07 keV and 968.90 keV) gamma ray transitions [24].

3.3.2. Self-Attenuation Factors

The high-resolution gamma-ray spectrometry is an effective method to analyze the activities of natural radionuclides in soils in order to assess their environmental impact. Low energy gamma rays have less penetrating ability and tend to interact more readily with matter, so, the gamma ray interaction is different for the sand samples and the aqueous standard solution [25].

The self-attenuation phenomenon depends on the sample composition, its density and the gamma emission energy.

Self-attenuation factors are needed because the sand samples from Camburi Beach have apparent densities in the range of 1.26 g/cm^3 to 2.35 g/cm^3 and the efficiency curve was obtained with a standard aqueous solution [26].

The apparent density was determined dividing the mass in grams of each sample by the flask volume of 100 mL.

The self-attenuation correction factors were experimentally determined by the Cutshall transmission method [25]. The method is based on measuring the transmission of gamma-ray through both the sand sample and an ultrapure water sample in the same geometry, with gamma transitions in the range of interest. Punctual standard gamma-ray sources of ^{60}Co, ^{133}Ba, ^{137}Cs and ^{152}Eu with gamma transitions ranging from 81 keV to 1408 keV were used, in order to cover the energies of the radionuclides used for the concentrations activities calculations.

3.4. Doses

Direct measurements of absorbed dose rates in air have been carried out in the last few decades in many countries of the world. The Scientific Committee on the Effects of Atomic Radiation (UNSCEAR) has evaluated the absorbed dose rates in air from these countries and several values have been reviewed based on new informations [1].

The International Commission on Radiological Protection (ICRP-60) [27], in order to maintain the radiological security of the population, recommends an annual effective dose value of 1 $mSv.y^{-1}$ for the general public.

In order to determine the level of public exposures on Camburi beach, the knowledge about absorbed dose rate in outdoor air (D) and the annual effective dose (*E*) plays an important role in radiation protection.

3.4.1. Absorbed Dose Rate in Outdoor Air (D)

The absorbed dose rate (D) in outdoor air at 1 m above the ground in $nGy.h^{-1}$ is calculated from equation (2) [1], as:

$$D = 0.462 A_{Ra} + 0.604 A_{Th} + 0.0417 A_K \tag{2}$$

where A_{Ra}, A_{Th} and A_K are, respectively, the activities concentrations of ^{226}Ra, ^{232}Th and ^{40}K, in $Bq.kg^{-1}$.

3.4.2. Annual Effective Dose (E)

The Annual Effective Dose (E) is calculated from equation (3) [9]:

$$E = D.t.0,7.10^{-6} \qquad (3)$$

where D is the absorbed dose rate in the air (nGy h^{-1}), t is the time in hours of exposure per year (24 hours X 365 days X fraction of the day exposed to radiation) and 0.7 is the conversion factor to unit $Sv.Gy^{-1}$.

The ICRP 60 [27] recommended value is 1 $mSv.y^{-1}$ for the general public.

4. Results and Discussion

4.1. Activities Concentrations of ^{226}Ra, ^{232}Th and ^{40}K in Camburi Beach Sands

The activities concentrations of ^{226}Ra, ^{232}Th and ^{40}K of sand samples from the collection points on Camburi beach during 2011 are presented in Figure 5, 6, 7, respectively. The activities concentrations of ^{226}Ra, ^{232}Th and ^{40}K were calculated by the equation (1).

As expected, due to the strong presence of monazite, the concentration of ^{232}Th is higher than the concentration of ^{226}Ra and ^{40}K.

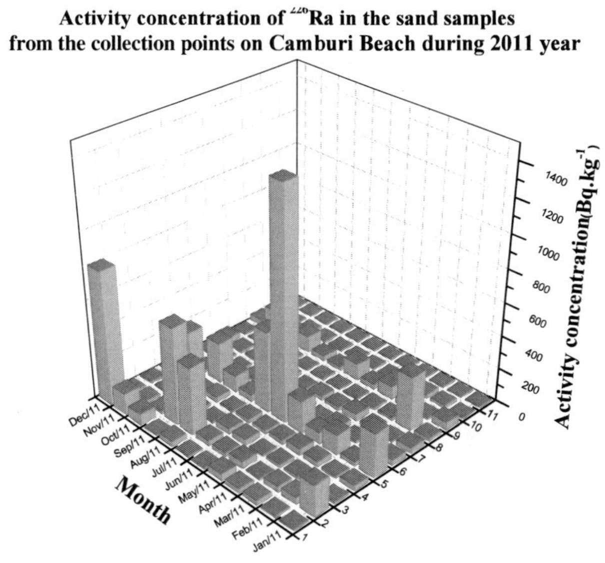

Figure 5. Activity concentration of ^{226}Ra in sand samples from the collection points (1-11) at Camburi beach, ES, Brazil, throughout the year 2011.

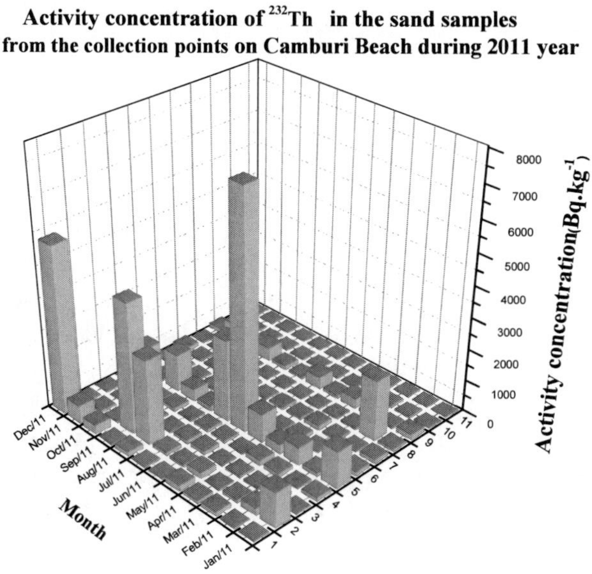

Figure 6. Activity concentration of ^{232}Th in sand samples from the collection points (1-11) at Camburi beach, ES, Brazil, throughout the year 2011.

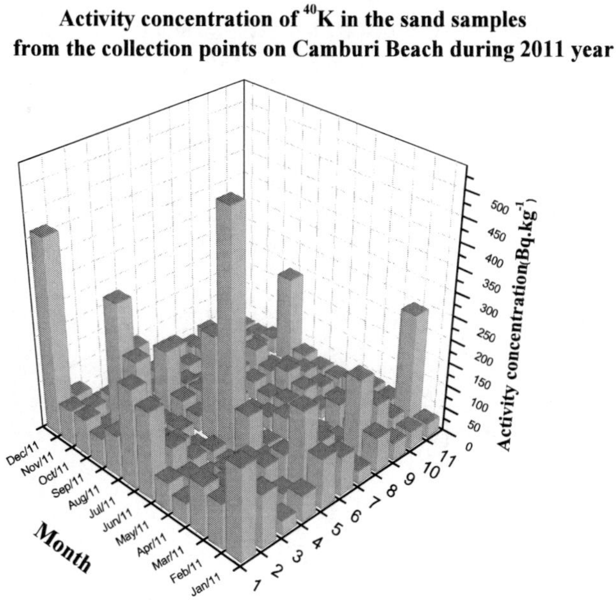

Figure 7. Activity concentration of ^{40}K in sand samples from the collection points (1-11) at Camburi beach, ES, Brazil, throughout the year 2011.

It was observed at point 5 the highest values of activity concentration of ^{226}Ra (1380 ± 62) $Bq.kg^{-1}$, ^{232}Th (7453 ± 274) $Bq.kg^{-1}$ and ^{40}K (504 ± 41) $Bq.kg^{-1}$ in June, followed by point 1 in December, with activity concentration of ^{226}Ra (806 ± 39) $Bq.kg^{-1}$, ^{232}Th (5283 ± 206) $Bq.kg^{-1}$ and ^{40}K (415 ± 36) $Bq.kg^{-1}$ and at point 2 in September, with activity concentration of ^{226}Ra (580 ± 26) $Bq.kg^{-1}$, ^{232}Th (4122 ± 150) $Bq.kg^{-1}$ and ^{40}K (320 ± 26) $Bq.kg^{-1}$.

The values found are in the range of literature values for similar sands (Table 1).

The activity concentration of ^{226}Ra, ^{232}Th and ^{40}K found in this work for the Camburi beach sand samples are closer to the values found in Guarapari beach sands (Table 1).

4.2. Activities Concentrations of ^{226}Ra, ^{232}Th and ^{40}K in Camburi Beach Sands and Average Temperatures of Vitória

The Figures 8, 9 and 10 respectively show graphs of the variation of the activity concentration of ^{226}Ra, ^{232}Th and ^{40}K in the sands collected in Camburi beach and monthly average temperature (ºC) -lighter and thicker line- in Vitória during the year 2011.

The average temperature data were extracted from charts of Espírito Santo Research Institute, Assistance and Rural Extension [29].

The monthly average temperature in Vitória in 2011 has maximum values in February, August and October and minimum values in July and September.

Table 1. Ranges of ^{226}Ra, ^{232}Th and ^{40}K activities concentrations of sand samples from some beaches of Brazil, India, Turkey, China and Egypt

	Brazil, Espírito Santo			Brazil	India	Turkey	China	Egypt
Activity Concentration ($Bq.kg^{-1}$)	Camburi beach	Vitória	Guarapari	Bahia	Kalpakkam	Marmara	Rizhao	El-Arish
^{226}Ra	4-1380	6-755	3-643	10-572	9.6-53	24-82	7,6-17.2	31-39
^{232}Th	9-7453	16-4155	9-7236	14-1735	37-163	15-57	7.8-25.1	10-15
^{40}K	6-504	29-377	14-669	25-62	210.1-607	255-877	883.4-1313.6	350-550
Reference	This work	[9]	[99]	[8]	[28]	[5]	[4]	[3]

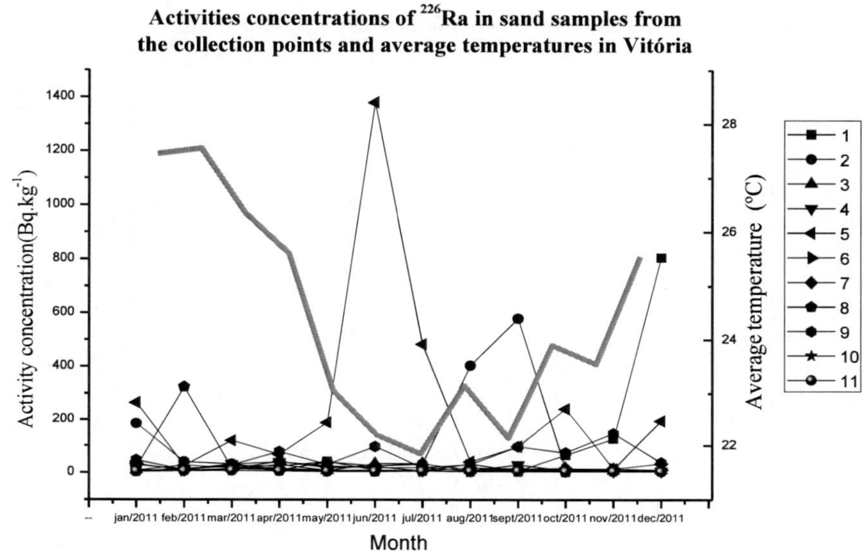

Figure 8. Activities concentrations of ^{226}Ra in sand samples from the collection points (1-11) and average temperatures in Vitória throughout the year 2011.

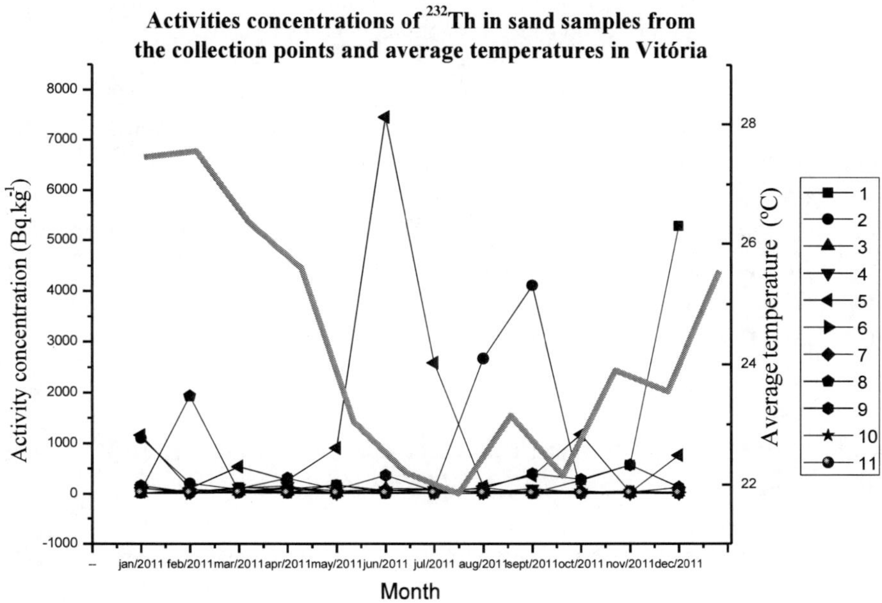

Figure 9. Activities concentrations of ^{232}Th in sand samples from the collection points (1-11) and average temperatures in Vitória throughout the year 2011.

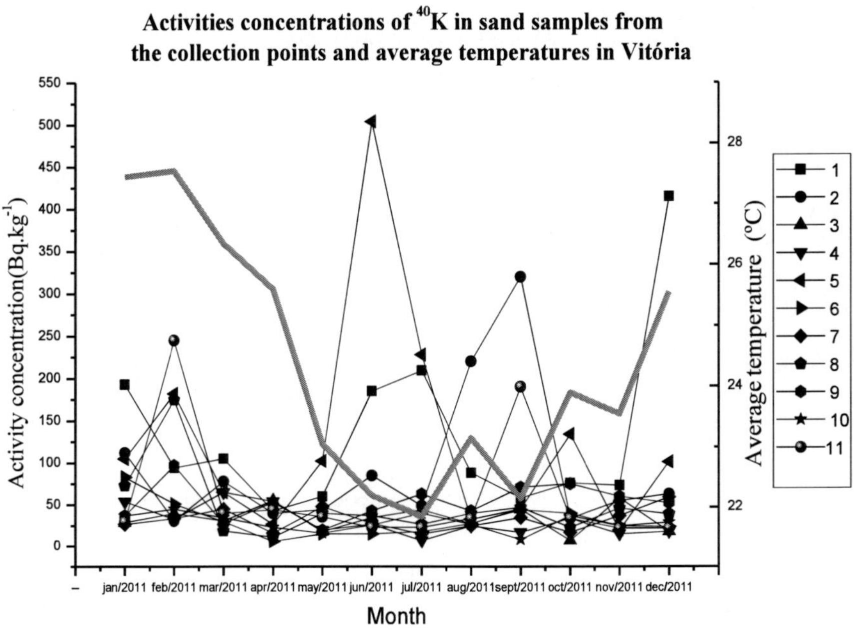

Figure 10. Activities concentrations of ^{40}K in sand samples from the collection points (1-11) and average temperatures in Vitória throughout the year 2011.

It is not an easy work to establish a reliable relationship between the values of the monthly average temperature in Vitória in 2011 and the activities concentrations of ^{226}Ra, ^{232}Th and ^{40}K of sand samples from Camburi beachnova, but in some cases it is possible. It is possible a relationship in April, when the low ^{226}Ra, ^{232}Th and ^{40}K activities concentrations in all the collection points can be related with the high average temperature. In June on point 5, the high ^{226}Ra, ^{232}Th and ^{40}K activities concentrations can be related with the low average temperature.

4.3. Activities Concentrations of ^{226}Ra, ^{232}Th and ^{40}K in Camburi Beach Sands and Rainfall Levels of Vitória

The Figures 11, 12 and 13 respectively show graphs of the variation of the activity concentration of ^{226}Ra, ^{232}Th and ^{40}K in the sands collected in Camburi beach and rainfall levels of Vitória (mm) -lighter and thicker line- during the year 2011. The rainfall levels data were extracted from charts of Espírito Santo Research Institute, Assistance and Rural Extension [29].

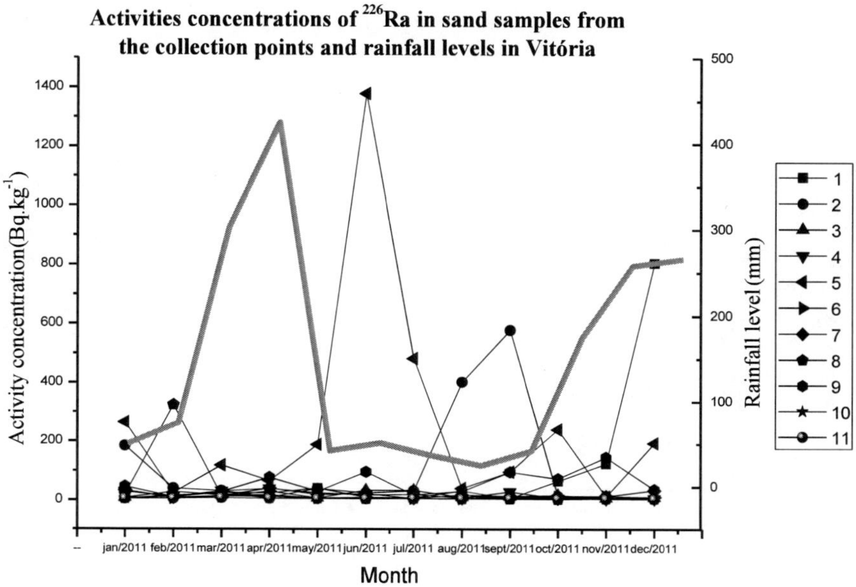

Figure 11. Activities concentrations of ^{226}Ra in sand samples from the collection points (1-11) and rainfall levels in Vitória throughout the year 2011.

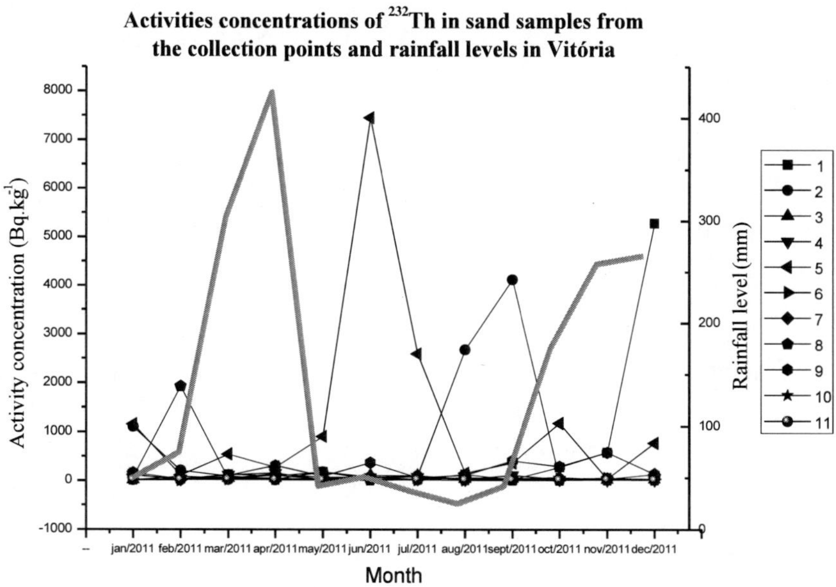

Figure 12. Activities concentrations of ^{232}Th in sand samples from the collection points (1-11) and rainfall levels in Vitória throughout the year 2011.

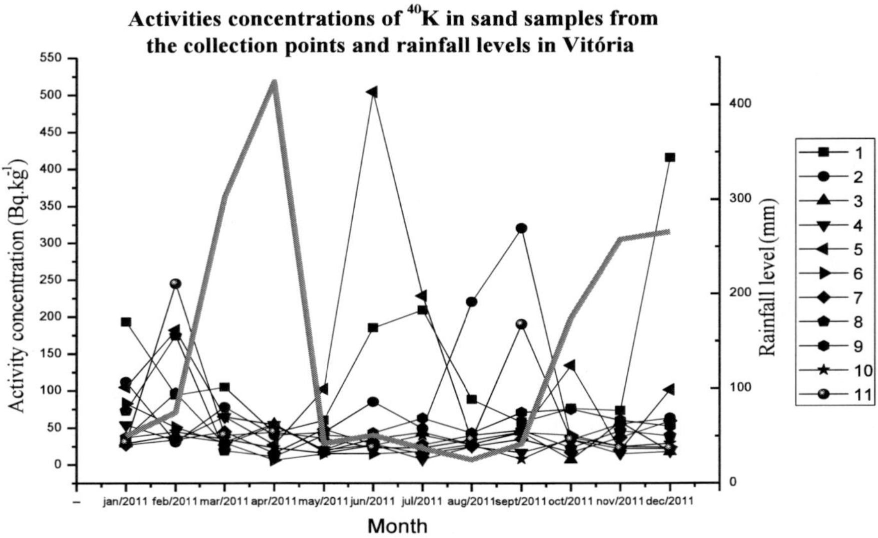

Figure 13. Activities concentrations of ^{40}K in sand samples from the collection points (1-11) and rainfall levels in Vitória throughout the year 2011.

A maximum value of rainfall level occurred during the month of April and another in December are shown on Figure 11,12 and 13.

The low activities concentrations of ^{226}Ra, ^{232}Th and ^{40}K in April and December could be related to the maximum value of the rainfall level, with the exception of point 1 in December. High rainfall can contribute in the dispersion of monazite, ilmenite, rutile and zircon in the sand [14].

The high ^{226}Ra, ^{232}Th and ^{40}K activities concentrations in June on point 5 could be related with the low rainfall level. Low rainfall can hamper the dispersion of monazite, ilmenite, rutile and zircon in the sand [14].

4.4. Activities Concentrations of ^{226}Ra, ^{232}Th and ^{40}K in Camburi Beach Sands and Variations of the Height of the Tides of Tubarão Seaport and Vitória Seaport

The Figure 14, 15 and 16 show graphs of the variation of the activity concentration of ^{226}Ra, ^{232}Th and ^{40}K in collection points and the variation of tidal heights (mm) -lighter and thicker line- on Tubarão Seaport and Vitória Seaport throughout the year 2011.

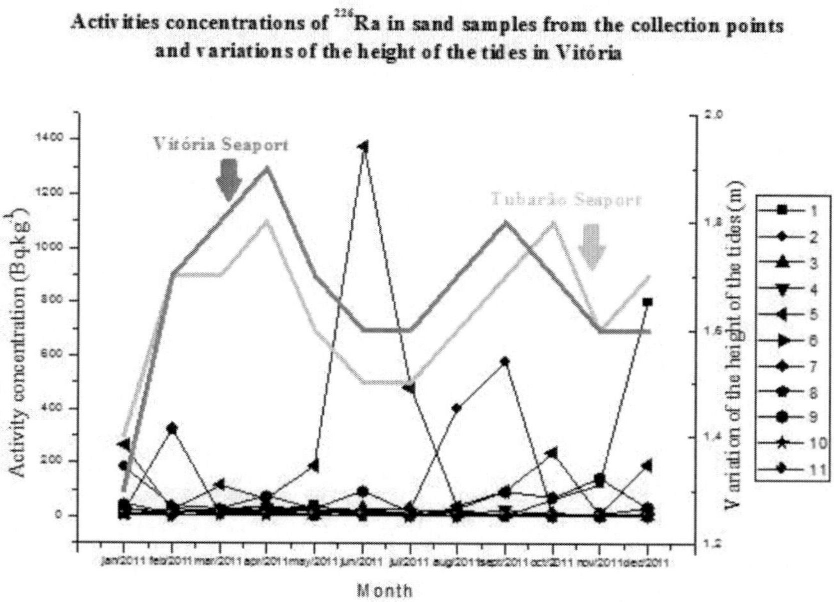

Figure 14. Activities concentrations of ^{226}Ra in sand samples from the collection points (1-11) and variations of the height of the tides in Vitória throughout the year 2011.

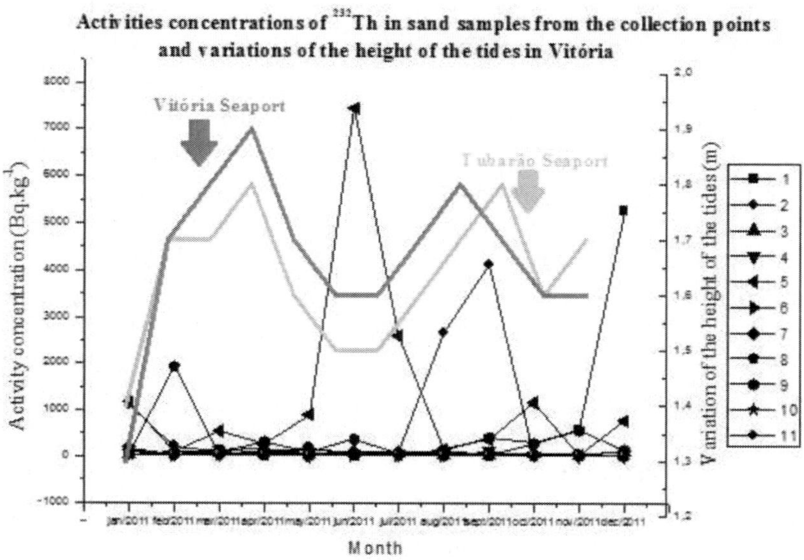

Figure 15. Activities concentrations of ^{232}Th in sand samples from the collection points (1-11) and variations of the height of the tides in Vitória throughout the year 2011.

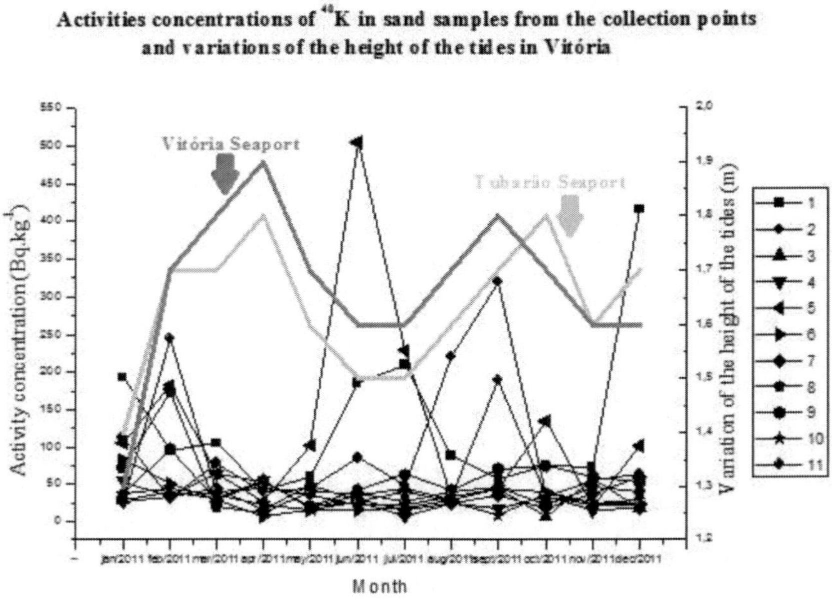

Figure 16. Activities concentrations of ^{40}K in sand samples from the collection points (1-11) and variations of the height of the tides in Vitória throughout the year 2011.

The values of the heights of the tides used were obtained from the National Oceanographic Data Bank [30].

The variation of the height of the tide has a maximum value occurred during the month of April in both seaports, one in September in the Vitória Seaport and other in October in the Tubarão Seaport. The Tubarão Seaport is closest to the Camburi beach, just after the Point 11, north of Camburi beach (Figure 2).

The low activities concentrations of ^{226}Ra, ^{232}Th and ^{40}K in all samples collected in April and October may be related to high tidal height variation found in these months at Tubarão Seaport.

The high variation of the height of the tides can contribute in the dispersion of monazite, ilmenite, rutile and zircon in the sands of the collection points [14].

The Socó Island (Figure 2) - in front of the collection point 5 – and the low variation of the height of the tides on June can be related with the high ^{226}Ra, ^{232}Th and ^{40}K activities concentrations on June in the collection point 5. These factors could contribute in the increase of the concentration of monazite, ilmenite, rutile and zircon in the sands on June in point 5 [14].

4.5. Doses

4.5.1. Absorbed Dose Rate in Outdoor Air (D) in Camburi Beach Sands

The absorbed dose rate (D) in outdoor air at 1 m above the ground (nGy h^{-1}) in sand samples collected in Camburi beach during 2011 is calculated from equation (2) and shown on Figure 17.

The absorbed dose rate (*D*) in outdoor air values laid from 9 nGy.h^{-1} to 5160 nGy.h^{-1}.

It was observed at point 5 the highest value of *D* (5160 nGy.h^{-1}) in June, followed by point 1 in December (3581 nGy.h^{-1}) and at point 2 in September (2771 nGy.h^{-1}).

The behavior of the variation of the absorbed dose rate (*D*) in outdoor air (Figure 17) is similar to the behavior of the variation of activities concentrations of the ^{226}Ra (Figure 5) and ^{232}Th (Figure 6) for the collection points throughout the year 2011.

The values found are in the range of literature values for similar sands (Table 2).

Absorbed dose rate (D) in outdoor air in the sand samples (1-11) from Camburi Beach during 2011

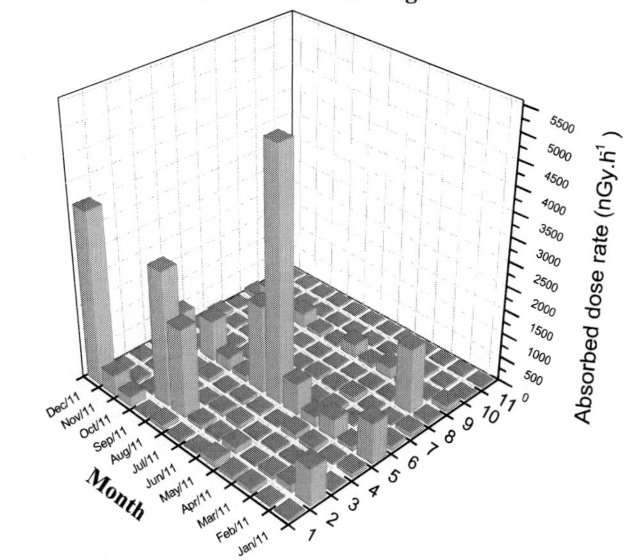

Figure 17. Absorbed dose rate (D) in outdoor air at 1 m above the ground in sand samples collected in Camburi beach during 2011.

Table 2. Ranges of absorbed dose rate (D) in outdoor air in the sand samples from the beaches of Brazil, India, Turkey, China and Egypt

	Brazil, Espírito Santo.			India	Turkey	China	Egypt
	Camburi beach	Vitória	Guarapari	Kalpakkam	Marmara	Rizhao	El-Arish
D $(nGy.h^{-1})$	9-5160	14-2874	6-4696	41.2-131.9	121	50.9-69.8	38.3
Reference	This work	[9]	[9]	[28]	[5]	[4]	[3]

The absorbed dose rate (*D*) in outdoor air found in this work for the Camburi beach sand samples are closer to the values found in Guarapari beach sands (Table 2).

4.5.2. Annual Effective Dose (E) in Camburi Beach Sands

To organize samples, a continuous identification number (ID) to according to the order of collection was considered. In January the sand samples were 1-11, February, 12-22, and so on, until December, with samples 122-132.

The annual effective dose of sand samples from the selected location at the most frequented region of Camburi beach is calculated by equation (3), considering the worst case scenario of a 24-hours daily exposure all year round (Figure 18). The value recommended by ICRP 60 [27] is 1 $mSv.y^{-1}$ (Figure 18) for the general public. The majority of the sand samples show values of *E* lower than 1 $mSv.y^{-1}$.

However, the annual effective doses established in Figure 18 are for a man who was in contact with the sand samples collected, 24 hours a day, during 1 year. This condition of 100% exposure (24 hours) to a sample of sand collected in Camburi Beach is the most extreme condition possible, an unlikely event.

The annual effective dose of sand samples from the selected location at the most frequented region of Camburi beach is calculated by equation (3), considering the scenario of one-hour daily exposure all year round (Figure 19).

Considering a 1-hour exposure per day for each sample in a year, only sample from point 5 collected in June exceeded 1 $mSv.y^{-1}$.

An assessment of the annual effective dose to a resident of Camburi who frequents the beach throughout the year and go to a single point on the beach all day with an exposure time of 1 hour per day, imagining that it is resting, sitting the sand is shown in Figure 20.

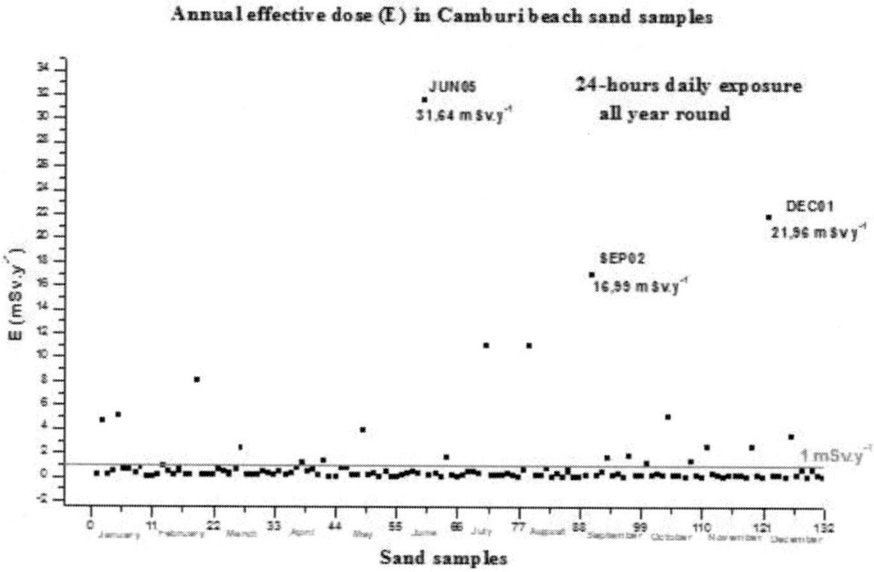

Figure 18. Annual effective dose (E) for sand samples collected in Camburi beach during 2011, 24-hours daily exposure all year round.

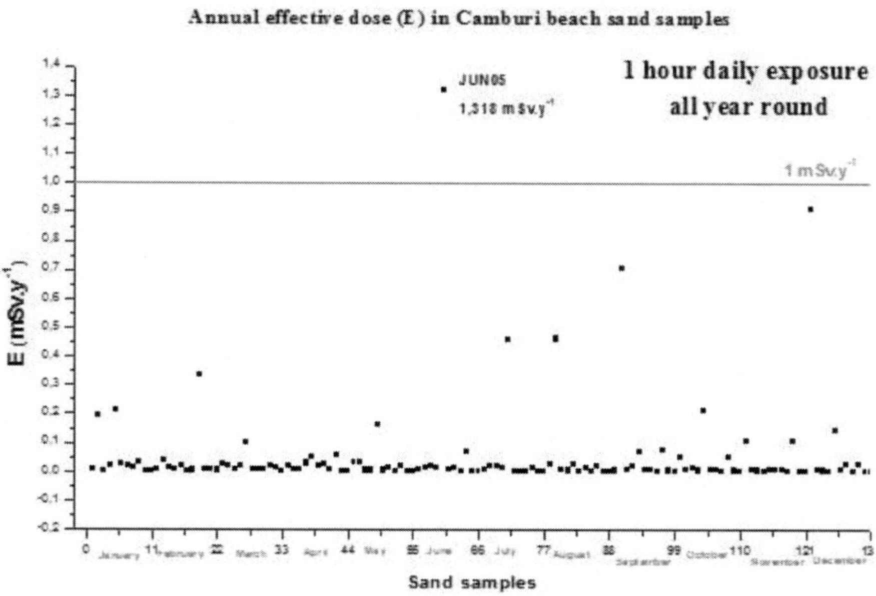

Figure 19. Annual effective dose (E) for sand samples collected in Camburi beach during 2011, 1-hour daily exposure all year round.

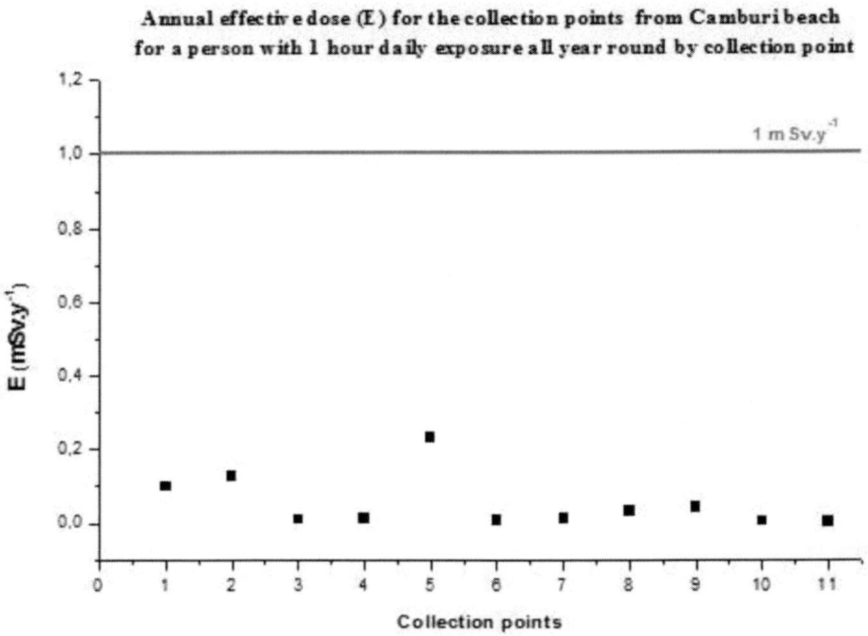

Figure 20. Annual effective dose (E) for the collection points from Camburi beach for a person with 1 hour daily exposure all year round.

The annual effective dose varied from 0.003 mSv.y^{-1} from point 11 to 0,232 mSv.y^{-1} from point 5. Therefore, the values found did not exceed the limit of 1 mSv.y^{-1}. An assessment of the annual effective dose was also taken for a tourist who goes to the beach on vacation in January, in July or in both (Figure 21), with an exposure time of 8 hours daily. The annual effective dose (E) for the collection points from Camburi beach for a tourist with 8 hour daily exposure on january, july or both laid from 0.004 mSv.y^{-1} to 0.446 mSv.y^{-1}, therefore the values found did not exceed the limit of 1 mSv.y^{-1}.

Conclusion

In this work, the influence of environmental aspects, as geology, average temperatures, variations of the height of the tides and the rainfall levels were checked with the variation of ^{226}Ra, ^{232}Th and ^{40}K activities concentrations during the year 2011 and radiation hazards indexes and annual effective dose were evaluated from the ^{226}Ra, ^{232}Th and ^{40}K sands activities concentrations.

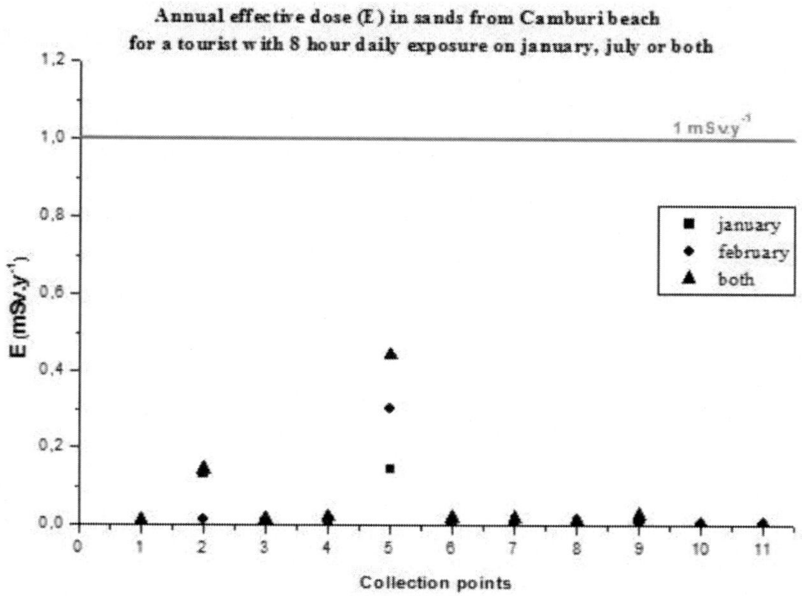

Figure 21. Annual effective dose (E) for the collection points from Camburi beach for a tourist with 8 hour daily exposure on january, july or both.

The results of the present work are related to locations at Camburi beach. Monthly samplings were done in 2011, covering all extension of the Camburi Beach mainland, with 11 samples locations.

It is possible to establish a temperature relationship in April, when the low ^{226}Ra, ^{232}Th and ^{40}K activities concentrations in all the collection points can be related with the high average temperature.

In June, at point 5, the high ^{226}Ra, ^{232}Th and ^{40}K activities concentrations can be related with the low average temperature. The low activities concentrations of ^{226}Ra, ^{232}Th and ^{40}K in April and December could be related to the maximum value of the rainfall level, with the exception of point 1 in December. A high rainfall can contribute in the dispersion of monazite, ilmenite, rutile and zircon in the sand [14].

The high ^{226}Ra, ^{232}Th and ^{40}K activities concentrations in June on point 5 could be related with the low rainfall level. Low rainfall can hamper the dispersion of monazite, ilmenite, rutile and zircon in the sand [14].

The low activities concentrations of ^{226}Ra, ^{232}Th and ^{40}K in all samples collected in April and October may be related to high tidal height variation found in these months at the Tubarão Seaport.

The high variation of the height of the tides can contribute in the dispersion of monazite, ilmenite, rutile and zircon in the sand [14]. The Socó Island (Figure 2) - in front of the collection point 5 – and the low variation of the height of the tides on June may be related with the high ^{226}Ra, ^{232}Th and ^{40}K activities concentrations on June in the collection point 5. These factors could contribute in the increase of the concentration of monazite, ilmenite, rutile and zircon in the sands on June in point 5 [14]. The constant dredging can change the concentrations of monazite, ilmenite, rutile and zircon in the sands. The condition of 100% exposure (24 hours) to a sample of sand collected in Camburi Beach is the most extreme condition possible, an unlikely event. Considering the exposure times reported in this work to a resident of Camburi and to a tourist, the sands showed values of *E* lower than 1 $mSv.y^{-1}$ do not show radiological risk to the population.

ACKNOWLEDGMENTS

The authors would like to thank the Institute for Energetic and Nuclear Researches (IPEN/SP) for the Master Degree opportunity, the Environmental Radiometric Laboratory of IPEN for the use of the facilities and the CNPq for the scholarship grant. One of the authors (Lívia F. Barros) would like to thank Mateus F. Barros, Elany Soraia F. Barros and Leonardo F. Santos for their assistance in sample collection.

REFERENCES

[1] UNSCEAR, *Sources and effects of ionizing radiation*, United Nations Scientific Committee on the Effects of Atomic Radiation, United Nations, New York, 2000.

[2] Uosif, M. A. M., El-Taher, A., Abbady, A. G. E. "Radiological significance of beach sand used for climatotherapy from Safaga, Egypt." *Radiation Protection Dosimetry.* 2008. vol. 131, 331-339.

[3] Eissa, H. S., Medhat, M. E., Said, S. A., Elmaghraby, E. K. "Radiation dose estimation of sand samples collect from different Egyptian beaches." *Radiation Protection Dosimetry.* 2010, vol. 147, 533-540.

[4] Xinwei, L., Zhang, X. "Measurement of natural radioactivity in beach sands from Rizhao bathing beach, China". *Radiation Protection Dosimetry.* 2008, vol. 130, 385-388.

[5] Cevik, U., Damla, N., Kobya, A. I., Celik, N., Celik, A., Van, A. A. "Assessment of natural radioactivity of sand used in Turkey." *Journal of Radiological Protection.* 2009, vol. 29, 61-74.

[6] Kannan, V., Rajan, M. P., Iyengar, M. A. R., Ramesh, R. "Distribution of natural and anthropogenic radionuclides in soil and beach sand samples of Kalpakkam (India) using hyper pure germanium (HPGe) gamma ray spectrometry." *Applied Radiation and Isotopes.* 2002, 109–119.

[7] Alam, M. N., Chowdhury, M. I., Kamal, M., Ghose, S., Islam, M. N., Mustafa, M. N., Miah, M. M. H., Ansary, M. M. "The ^{226}Ra, ^{232}Th and ^{40}K activities in beach sandminerals and beach soils of Cox's Bazar, Bangladesh." *Journal of Environmental Radioactivity.* 1998, vol. 46, 243-250.

[8] Veiga, R., Sanches, N., Anjos, R. M., Macari, K., Bastos, J., Iguatemy, M., Aguiar, J. G., Santos, A. M. A., Mosquera, B., Carvalho, C., Filho, M. B., Umisedo, N. K. "Measurement of natural radioactivity in Brazilian beach sands." *Radiation Measurements.* 2006, vol. 41, 189-196.

[9] Aquino, R. R. "Avaliação da radioatividade natural em areias das Praias da Grande Vitória - Espírito Santo". *Thesis (Master Degree) - Instituto de Pesquisas Energéticas e Nucleares-IPEN-CNEN/SP,* São Paulo, 2010, pp. 79 <http://www.teses.usp.br>.

[10] Andrade, R. H. P. National Department of Mineral Production. Mineral Summary 2010. Rare Earths. Mato Grosso do Sul, 2010. <https:// sistemas.dnpm.gov.br/publicacao/mostra_imagem.asp?IDBancoArquivo Arquivo=5492>

[11] Eisenbud, M. *Environmental Radioactivity.* 2nd ed. Academic Press, Orlando, 1987.

[12] Moura, J. C. "Estudo da Variação Sazonal dos Níveis de Radiação Gama da Praia da Areia Preta, Guarapari, Espírito Santo, Brasil: Radiometria e Análise de Risco." *Thesis (Master Degree) - Universidade Federal do Espírito Santo,* Espírito Santo, 2003.

[13] Lauria, D. C., Rochedo, E. R. R. The legacy of monazite processing in Brazil. *Radiation Protection Dosimetry.* 2005. vol. 114, n. 4, 546-550.

[14] Barros, L. F. "Avaliação da variação da radioatividade natural em areias da Praia de Camburi-Vitória-Espírito Santo com fatores climatológicos e

geológicos da região." *Thesis (Master Degree) - Instituto de Pesquisas Energéticas e Nucleares-IPEN-CNEN/SP*, São Paulo, 2013, pp. 125.

[15] Barros, L. F., Pecequilo, B. R. S. "Radiation Hazards Evaluation for selected sand samples from Camburi Beach, Vitória, Espírito Santo, Brazil. *2013 International Nuclear Atlantic Conference,* Recife, Brazil, November 24-29, 2013.

[16] Barros, L. F., Pecequilo, B. R. S. "Evaluation of Ra, Th, K and radium equivalent activity in sand samples from Camburi Beach, Vitória, Espírito Santo, Brazil." *IX Latin American IRPA Regional Congress on Radiation Protection and Safety,* Rio de Janeiro, Brazil, April 15-19, 2013.

[17] Barros, L. F., Pecequilo, B. R. S., Aquino, R. R. "Gamma radiation measurements in select sand samples from Camburi Beach - Vitória, Espírito Santo, Brazil: Preliminary results." *2011 International Nuclear Atlantic Conference,* Belo Horizonte, Brazil, October 24-28, 2011.

[18] Sabaini, R. S. "Determinação da taxa de transporte de sedimentos na zona de espraiamento de um trecho da praia de Camburi." *Thesis (Master Degree) - Universidade Federal do Espírito Santo*, Espírito Santo, 2008.

[19] Sperandio, E. F. "Radioatividade Natural e Composição das Areias da Praia de Camburi (Vitória–ES, Brasil)." *Thesis (Master Degree) - Universidade Federal do Espírito Santo*, Espírito Santo, 2011.

[20] "Camburi beach map". <https://maps.google.com.br/>, 2013.

[21] "Brazil map (in black Espírito Santo State)". <http://www.guianet.com.br/es/mapaes.htm>, 2013.

[22] "Espírito Santo State map". <http://www.alunosonline.com.br/geografia/espirito-santo.html>, 2013.

[23] Knoll, G. F. *Radiation Detection and Measurement*, Third Printing, New York, United States, 1999.

[24] ORTEC, *InterWinnerTM6.0 MCA Emulation, Data Acquisition and Analysis Software For Gamma and Alpha Spectroscopy*, TOMCOM Software Ltd., 2004.

[25] Cutshall, N. H., et al. "Direct analysis of ^{210}Pb in sediment samples: self-absorption corrections," *Nuclear Instruments and Methods in Physics Research.* 1983. vol. 206, 309-312.

[26] Barros, L. F., Pecequilo, B. R. S. "Self-Attenuation Factors in gamma-ray spectrometry of select sand samples from Camburi Beach, Vitória, Espírito Santo, Brazil.", *Radiation Physics and Chemistry*. 2014. vol. 95, 339–341. <http://dx.doi.org/10.1016/j.radphyschem.2012.12.031>.

[27] ICRP 60. International Commission on Radiological Protection, Recommendation of the ICRP, publication 60, Ed. Pergamosn Press, Oxford, 1990.

[28] Sowmya, M., Senthilkumar, B., Seshan, B. R. R., Hariharan, G., Purvaja, R., Ramkumar, S., Ramesh, R. “Some natural radioactivity and associated dose rates in soil samples from Kalpakkam, South India.” *Radiation Protection Dosimetry*. 2010, vol. 141, 239-247.

[29] INCAPER. Espírito Santo Research Institute, Assistance and Rural Extension. < http://hidrometeorologia.incaper.es.gov.br/?pagina=vitoria _sh >. 2013.

[30] BNDO, National Oceanographic Data Bank. http://www.mar.mil.br/dhn/ chm/tabuas/index.htm>. 2012.

In: Beaches
Editor: Margaret Cessa

ISBN: 978-1-63117-239-7

Chapter 5

NUISANCE ALGAL BLOOMS

E. Pip*

University of Winnipeg, Winnipeg, Manitoba, Canada

ABSTRACT

Nuisance algae at beaches are responsible for aesthetic concerns, economic losses and health risks due to production of potent toxins and association with a variety of undesirable bacteria.

Nutrient enrichment with phosphorus, as well as nitrogen in N-limited systems, contributes to eutrophication and development of problem algal blooms. Management of nuisance algae at beaches includes reduction of nutrient inputs from the watershed, reduction of shoreline nutrient sources and shoreline erosion, prompt removal of algal accumulations washed up on shore, education and involvement of the public regarding ways to reduce the problem as well as awareness of health risks associated with algal blooms, and monitoring and timely posting of beach warnings. Use of algaecides is not recommended due to adverse effects on an already stressed ecosystem.

Algal blooms are a major cause of beach closures and fouling of recreational freshwaters. Global in occurrence, the problem is associated with increasing eutrophication, accelerated by a variety of human activities and land uses which promote nutrient escape into surface waters: sewage and

* e.pip@uwinnipeg.ca.

storm water outfalls, agricultural and cosmetic fertilizers, intensive livestock production, industry, recreational activities, deforestation, mining, and wetland destruction (Pip 2005). Runoff and erosion facilitate the transport of nutrients to surface waters from innumerable point sources within the watershed system, while the high-energy environment at beaches and exposed shorelines contributes to resuspension and recycling of the resident nutrient pool.

The most visible indicator of excessive nutrient loading is the appearance of nuisance algal blooms. Freshwater algal communities are composed of a variety of eukaryotic and prokaryotic algae, which in a stable ecosystem follow a regular seasonal progression and are neither conspicuous nor deleterious to the functioning of the system as a whole. During bloom conditions, planktonic and periphytic algal populations are overtaken by a few species which proliferate to nuisance proportions. Bloom composition may be various: eukaryotes such as chlorophytes, chrysophytes, dinoflagellates and euglenoids, or prokaryotic cyanobacteria may erupt under their respective favorable environmental circumstances. Blooms may follow seasonal patterns year to year, or show erratic unpredictable fluctuations. The photic zone becomes compressed as crowding and self-shading become more severe, and subsurface strata suffer oxygen depletion due to reductions in photosynthesis by macrophytes and deeper algal communities. Decomposition of the bloom biomass results in elevated oxygen demand as well as ammonia release, which may result in fish kills (Paerl 1988), microbial production of noxious anaerobic metabolites such as methane and toxic hydrogen sulfide, and reduction in biodiversity of the aquatic ecosystem.

Nuisance algal blooms are responsible for significant economic losses, which in the U.S. alone are estimated in the billions of dollars annually (Vazquez-Coriano 2011). Aesthetic concerns affect property values, tourism and local recreation-based businesses, and generate unfavorable publicity. Algal mats foul fishing nets, and tend to wash ashore and decompose, soiling beaches and shorelines, and are often associated with objectionable odors. At least 200 algal odoriferous metabolites have been found (Watson 1999 *in* Watson 2003). Cyanobacteria may produce compounds such as geosmin, 2-methylisoborneol, and isopropylthiols (Watson 2003, Watson et al. 2008). Chrysophytes, haptophytes, cryptophytes, dinoflagellates, as well as planktonic cyanobacteria are associated with volatile polyunsaturated fatty acid derivatives, while chlorophytes may produce aliphatic hydrocarbons (Watson 2003, Watson et al. 2008). Nor-carotenoids are produced by a variety of prokaryotic and eukaryotic algae (Watson 2003). Many of these compounds may have biological activity as grazer and parasite deterrents, allelogens, pheromones, kairomones or other roles (Watson 2003). Senescence and decomposition of algal biomass is accompanied by increased release of intracellular metabolites (Watson 2003), as well as additional odors arising from volatile degradation products generated by associated microbiota such as sulphate-reducing bacteria harbored in the algal mats (Olapade et al. 2006).

Blooms at recreational beaches are accompanied by various risks to human health. A diverse assemblage of heterotrophic bacteria is associated with the organic-rich environment of dense algal populations; some of these bacteria include human pathogens (Dziallas and Grossart 2011). Mats of the filamentous green alga *Cladophora* along beaches in the Great Lakes region have been shown to harbor and enhance the survival of fecal bacteria such as *Escherichia coli* and enterococci, which may be ingested by swimmers or children playing near the shore (Byappanahalli et al. 2006, Englebert et al. 2008). Dried algal mats may provide sources for environmental reinoculation by fecal bacteria upon rehydration even after several months (Whitman et al. 2003).

However the most dramatic noxious blooms are the cyanobacteria or bluegreen algae, which also present the most frequent health hazards at freshwater beaches. These prokaryotes have the ability to reproduce rapidly, outcompeting other algae by aggressive nutrient uptake. They produce intracellular gas vesicles (Humphries and Lyne 1988), which allow them to rise to the top of the photic zone and form floating mats that monopolize the light (Paerl 1988). Additional buoyancy may be conferred by bubbles of photosynthetic oxygen which remain trapped in the mucilaginous matrix of cell agglomerations; benthic algal slimes may break free and wash ashore via this phenomenon (Chorus et al. 2000).

A number of bloom-forming cyanobacterial genera are capable of producing a variety of potent cyanotoxins, many of which have yet to be identified (Blaha et al. 2009, Rastogi and Sinha 2009). Various functions have been proposed for cyanotoxins: anti-herbivory strategies (e.g., Lampert 1981), allelopathy against other algal competitors (Kearns and Hunter 2001), protection against oxidative stress from reactive oxygen species generated by ultraviolet light (Paerl and Otten 2013), and agents for increasing the availability of phosphorus for uptake (Leao et al. 2012). While the majority of many toxins is held within the living cells, toxin release occurs upon cell death or damage (Lahti et al. 1997, Watson et al. 2008). No consistent relationship appears to exist between odor and toxin level (Watson et al. 2008).

Toxins may be ingested while swimming, absorbed through contact, or inhaled (Stewart et al. 2006) as volatile agents from algal mats on beaches, or in spray aerosols from wave action, splashing, boating, water skiing, or using personal watercraft (Backer et al. 2008), or showering (Vazquez-Coriano 2011). Water trapped in bathing suits may prolong contact time (Chorus et al. 2000). Algal toxins elicit a wide spectrum of reactions, from mild and self-limiting to acute, including gastroenteritis, headaches, fever, respiratory symptoms and dermatitis (see Stewart et al. 2006), related to route of exposure. Numerous deaths of dogs, livestock and wildlife have been reported at beaches, as animals may drink large quantities of contaminated water. Many cases of human illness have occurred after swimming at beaches infested with algal blooms (e.g., Dillenberg and Dehnel 1960, Chorus et al. 2000, Blaha et al. 2009). Reliable estimates of incidence of illness due to cyanobacterial exposure are not available, and it is believed that the majority of such illnesses are un- or misdiagnosed and unreported (e.g., Yoo et al. 1995 *in* Chorus et al. 2000).

Cyanotoxins can be grouped according to their primary mode of action: neurotoxins, hepatotoxins, and irritant toxins, which are inflammatory agents and include dermatotoxins, cytotoxins and lipopolysaccharide endotoxins (Wiegand and Pflugmacher 2005, Blaha et al. 2009). The type or combination of types of toxins produced are strain specific (Zurawell et al. 2013); thus a given cyanobacterial species may have toxic and nontoxic genetic strains, which cannot be distinguished morphologically (Davis et al. 2009). For example, 5 – 100% of *Microcystis* in Lake Erie blooms may consist of toxic strains (Rinta-Kanto et al. 2005). However toxic strains may not always produce toxins (Vazquez-Coriano 2011).

Cyanobacterial neurotoxins cause cramping, rapid paralysis and death in vertebrates (Carmichael 1994). Known neurotoxins include anatoxins, which include secondary amines and an ester of a cyclic hydroxyguanidine molecule (Neilan et al. 2013). Anatoxins are either not hydrolyzed by acetylcholinesterase, or bind to it irreversibly (Carmichael 1994). Anatoxins are known in certain species of *Anabaena, Aphanizomenon* and *Oscilatoria* (see review by Zurawell et al. 2013). Saxitoxins are a class of alkaloids which in freshwaters have been reported for *Aphanizomenon, Anabaena* and *Lyngbya*, among others (Neilan et al. 2013). Jamaicamides and kalkitoxin are lipopeptide neurotoxins which have been studied in *Lyngbya* (Rastogi and Sinha 2009, Neilan et al. 2013).

Cyanobacterial hepatotoxins are reportedly more common than neurotoxins (Chorus et al 2000). They target hepatocytes, cause blood to pool in the liver, with eventual death from circulatory collapse; at sublethal concentrations hepatotoxins cause and promote liver tumors (hepatocellular carcinoma) (e.g., Yu 1995, Falconer 1999, Blaha et al. 2009). This toxin group includes the cyclic peptides microcystins (heptapeptides) and nodularins (pentapeptides), and the alkaloid cylindrospermopsin (Chorus et al. 2000). Of these, the microcystins (MCs) have been the most widely studied. Genera which have been reported to produce MCs include *Anabaena, Aphanizomenon, Microcystis, Nostoc* and others (Chorus 2001). The mode of action, structure and known analogues of MCs have been reviewed by Zurawell et al (2013). At least 90 isoforms of MCs are thus far known (see Neilan et al. 2013); of these, MC-LR is one of the most common variants (Dawson 1998). MCs may impair and bioaccumulate in various components of the aquatic ecosystem (see the reviews of de Figueiredo (2004), Zurawell et al. (2013)), and may adversely alter the composition and species richness of invertebrate communities, with eventual broader consequences for the aquatic ecosystem (Lance et al. 2010).

Despite the increasing concerns posed by cyanobacterial blooms, the only toxin guideline currently established by the World Health Organization is 1.0 µg/L of MC-LR for drinking water. While many jurisdictions have established their own guidelines for beach waters, the problem is compounded by the impracticality of monitoring the many toxin isoforms as well as variety of toxin types (Vazquez-Coriano 2011).

A number of environmental factors promote cyanobacterial growth and dominance. Both phosphorus and nitrogen have been identified as primary nutrients in the promotion of cyanobacterial blooms (e.g., Rapala et al. 1997, Te and Gin 2011, see review of Lewis and Wurtsbaugh 2008), and high concentrations of N and P have been correlated with MCs (e.g., Rapala et al. 1997, Vezie et al. 2002, Te and Gin 2011, Pip and Bowman submitted). While total N and P are often used in trophic assessment, potentially bioavailable nutrients are represented primarily by the dissolved inorganic fractions (Pick and Lean 1987, Sonzogni et al. 1982) and therefore have the most immediate effects on algal growth. A number of workers have reported that MC production, and especially MC-LR, increases with increasing limitation of P (see Zurawell et al. 2013, Oh et al. 2000, Pip and Bowman submitted). Abundance of toxigenic *Microcystis* has been correlated with N and P (Te and Gin 2011), and toxic strains may have higher N and P requirements than nontoxic strains (Vezie et al. 2002).

In eutrophic phytoplankton communities, particularly those dominated by a small number of toxic strains, MCs can often be correlated with chl *a* (Oh et al. 2001, Pip and Bowman submitted). In eutrophic conditions, toxigenic strains comprise the majority of bloom phytoplankton (Paerl and Otten 2013).

However phytoplankton composition, and hence relative toxicity, often varies spatially as well as year to year in a given lake (Carmichael and Gorham 1981, Pip and Bowman submitted, and a number of studies have shown no consistent relationship between algal biomass and MCs (e.g., Sinang et al. 2013). Furthermore, percentage of toxigenic *Microcystis* may not be correlated with MC concentration (Te and Gin 2011).

Nutrient management remains the single most important focus of algal bloom control. Two approaches exist (see review of Lewis and Wurtsbaugh 2008): reduction of both nitrogen and phosphorus (e.g., Jeppesen et al. 2007), or reduction of phosphorus alone (e.g., Schindler et al. 2008). The latter strategy is based on resource-ratio theory (Downing et al. 2001) and targets the N/P ratio, where P is considered to be the primary limiting factor. At low N/P ratios, cyanobacteria with heterocysts are favored, since they are able to fix atmospheric nitrogen and thus may circumvent nitrogen limitation (Smith 1983, Schindler et al. 2008). According to the 'Redfield ratio' (Correll 1998), nitrogen limitation occurs below a total N/P ratio of 7:1, in terms of mass proportion. Pawlik-Skowronska et al. (2013) reported an inverse correlation between the dissolved N/P ratio and cyanobacterial biomass. A Canadian survey indicated a higher risk of MCs at low (total) N/P ratios (Orihel et al. 2012). In an extreme proposal, Levich (1996) even suggested that nitrogen be *added* to eutrophic systems to manipulate the N/P ratio, to replace cyanobacterial blooms with blooms of green algae, which are less toxic and more readily utilized by grazers.

According to Downing et al. (2001), the premise that low N/P ratios result in noxious and often toxic blooms is inaccurate. The latter authors concluded from a review of the literature that cyanobacterial blooms are more consistently correlated with P and N concentrations than the N/P ratio. Stanley et al. (2003) and Lewis and Wurtsbaugh (2008) contended that algal biomass is related to both N and P, based on evidence that algal blooms are at least as likely to be limited by N as by P. According to the latter authors, while low N/P ratios may favor cyanobacterial N fixers, the latter do not completely offset N limitation.

Indeed, *Microcystis*, which is a major bloom producer, is non-nitrogen fixing, and may show more growth and produce more toxin under N rich conditions (Watanabe and Oishi 1985, Orr and Jones 1998, Sivonen and Jones 1999 *in* Te and Gin 2011, Gobler et al. 2007, Davis et al. 2009). At high N levels, toxic strains of *Microcystis* may also outgrow nontoxic strains (Vezie et al. 2002).

All other (i.e., non nitrogen fixing) prokaryotic and eukaryotic algae benefit from N loadings, thus nuisance blooms, albeit of different composition, may still occur. Stanley et al. (2003) reported significant correlations between chl *a* and both total P and N. Pip and Bowman (submitted) found that in a high bloom year, chl *a* (representing total phytoplankton) was strongly correlated with dissolved P as well as nitrate, but not with N/P ratios, while chl *a* was correlated with dissolved P and N vertically and horizontally within a Precambrian Shield lake (Pip 1988). Pip and Bowman (submitted) also found that MCs were correlated with dissolved P, followed by nitrate, but correlation with the N/P ratio was inverse and much less significant than for the P and nitrate concentrations. Furthermore, MC concentrations lagged significantly behind changes in chl *a*, dissolved inorganic P, and nitrate-N, and less so (inversely) behind inorganic N/P ratios (Pip and Allegro 2010, Pip and Bowman submitted).

Pawlik-Skowronska et al. (2013) reported that during low dissolved N/P ratios, anatoxin-a production was favored compared to MCs. Thus while bloom toxicity is usually monitored in terms of MCs, it is important to keep in mind that other toxins are likely produced at the same time, and lower MC levels may thus not necessarily reflect low bloom toxicity. Nutrient reduction strategies may also not noticeably address chrysophytes, and may indeed aggravate them in eutrophic systems (Watson et al. 2008).

In terms of practical feasibility, Lewis and Wurtsbaugh (2008) suggested that regardless of whether N or P limitation is present, reduction of P can benefit a polluted system either by intensifying existing P limitation, or inducing P limitation where N limitation already exists. However the latter authors conceded that N reduction may be necessary in some N limited lakes.

While nutrients appear to play the greatest role, other environmental factors may also contribute to bloom development. Water pH has been identified as positively correlated with MCs (Wicks and Thiel 1990, Te and Gin 2011, Pip and Bowman submitted). Higher pH is normally associated with the greater rates of photosynthesis of algal blooms (see Te and Gin 2011), although pH may also play more subtle roles, for example in the modulation of microbial activity and P release from sediments by the activation of different alkaline phosphatases (see Pip and Allegro 2010).

Light is an important determinant of productivity and is correlated with chl *a* as well as temperature in the water column (e.g., Pip 1988). According to Utkilen and Gjolme (1992), light appears to affect MC production in *M. aeruginosa* when light intensity is limiting, otherwise maximum MC content is expected at the water surface, but in natural environments, light intensity may have a relatively minor effect on MCs (Bottcher et al. 2001). While cyanobacteria produce a number of "sunscreen" molecules that absorb ultraviolet light (Leao et al. 2012), high-irradiance environments may select for toxigenic strains (Paerl and Otten 2013). Ultraviolet light (particularly UVB), as well as algal photosynthesis, produce reactive oxygen species such as superoxide and hydrogen peroxide, which may favor toxigenic strains, possibly because MC may serve a protective function against oxidative stress (Ding et al. 2013, Paerl and Otten 2013).

The effect of temperature on MCs varies with genetic strain of various cyanobacteria (Zurawell et al. 2013). McQueen and Lean (1987) found that temperature showed a strong correlation with percent bluegreen algal composition. MCs generally increase with increasing temperature (Sivonen 1990, Pip and Bowman submitted), and the effect of temperature is particularly enhanced with N and P enrichment (Davis et al. 2009). However the type of MC produced may differ with temperature: Rapala et al. (1997) found that *Anabaena* MC-LR and MC-RR were favored at lower and higher temperatures respectively. Rapala and Sivonen (1998) further suggested that MC-LR and MC-RR are controlled largely by light and temperature respectively. Toxigenic strains are usually favored when temperature, light and photoperiod are greatest (Davis et al. 2009, Paerl and Otten 2013). However MC eruptions are not limited to the warmest time of year (Pip and Bowman submitted). With expected global warming, as well as intensifying anthropogenic eutrophication, increased incidence of toxic blooms is projected (Davis et al. 2009, Te and Gin 2011).

Iron and zinc are essential for cyanobacterial growth. Lukac and Aegerter (1993) found that concentrations of zinc were positively, and iron negatively, related to MCs, although other conflicting results have been attributed to differences among strains (see Zurawell et al. 2013). Under iron-limiting conditions, some cyanobacteria produce hydroxamate or catechol siderophores, which can scavenge and transport iron into the cells (Brown and Trick 1992). Some siderophores can also bind copper so that it is not assimilated, reducing its toxicity (Leao et al. 2012).

Other factors that may affect MC production include allelopathic compounds, and presence of zooplankton (Jang et al. 2008). Ecosystem alteration may alter bloom composition. For example, despite abatement of the green alga *Cladophora* by P management, invasion of the Great Lakes by zebra mussels (*Dreissena polymorpha*) has enhanced water clarity which has increased the depth of the photic zone (Auer et al. 2010), and possibly altered the pattern of P cycling (Bootsma 2007 *in* Englebert et al 2008), augmenting productivity of this green alga to problem proportions again.

MCs can persist in water for significant amounts of time after the bloom has declined and is no longer visible (Jones and Orr 1994, Lahti et al. 1997), and may remain in dried algal mats cast up on beaches (Chorus et al. 2000). MCs may undergo direct photodegradation at a wide range of wavelengths from photosynthetically active radiation to UV-A and UV-B (Wormer et al. 2010); decomposition is accelerated when cyanobacterial pigments are also present (Tsuji et al. 1994). Photodegradation may be prolonged by MC adsorption onto fine sediment and suspended particles (Morris et al. 2000). However eventual biodegradation by heterotrophic bacteria associated with algal blooms appears to be the primary route of toxin disposal (Dziallas and Grossart 2011), although several weeks may be required to achieve undetectable levels of toxin (Jones and Orr 1994).

Beaches provide foci where algal mats may become concentrated by wind, and MC concentrations may reach acute toxicity levels (Murphy et al. 2003). The gelatinous consistency of scums cast up in the swash zone may be attractive to children playing (Chorus et al. 2000), and adults might not be aware of the risks. Additional concerns resulting from association of algal mats with fecal bacteria have led to recommendations of mat removal for aesthetic and beach water quality reasons (Heuvel et al. 2010). However certain methods of beach grooming may shelter and reintroduce bacteria to beach foreshore sand and water; Kinzelman et al. (2003) found that mechanical groomers resulted in significantly higher *E. coli* counts compared to hand raking. Another temporary measure that has been suggested is to prevent mats from washing up on shore by the use of offshore booms or surface barriers (Chorus et al. 2000).

While substantial nutrient loading associated with runoff, leaching and erosion originates within the watershed, direct inputs and shoreline erosion are also significant contributory sources. The erosive forces of wave action, storms, water level fluctuations and surges, and ice scouring can be mitigated only partially by devices such as breakwaters, groins and rip-rap. At beaches and recreational areas, however, various human activities serve to accelerate erosion; these include foot traffic, digging, undermining banks, destruction of nearshore macrophyte communities and removal of stabilizing beachfront vegetation (e.g., Liddle and Scorgie 1980). High concentrations of visitors, particularly on hot days, are associated with direct nutrient influx associated with waste from humans (e.g., Liddle and Scorgie 1980), as well as from pets and wildlife such as gulls (Kinzelman et al. 2003). Inputs from percolating sewage and lawn and garden fertilizers from cottage and recreational properties in close proximity to the water are especially problematic in sandy soils. Such anthropogenic sources provide the most immediate and undiluted nearshore impacts, and may become initiating foci for algal blooms (Izydorczyk et al. 2008, Pip and Allegro 2010).

At beaches that are exposed to wave action, the effects of erosion may be aggravated by sediment resuspension. Sediment erosion rates are linked to bulk density of sediments (Jepsen et al. 1997). Impacts from wind-induced events may last from hours to days (Blom et al. 1994). Turbulence may contribute to the recycling of N and P from sediments (e.g., Hamilton and Mitchell, 1997, Hansen et al. 1997, Schallenberg and Burns 2004), although, depending on their nature, resuspended sediment particles may also adsorb inorganic P (Hansen et al. 1997). Availability of P associated with resuspension is affected by P concentration, pH, frequency of resuspension and other factors (see Schallenberg and Burns 2004). Resuspended particles may affect light attenuation, which may yield a small net reduction in local phytoplankton production (Hellstrom 1991), have little effect on light limitation (Schallenberg and Burns 2004), or increase productivity through the removal of surface photoinhibition (Ogilvie and Mitchell 1998). However increases in chl *a* may be significant as turbulence resuspends algae from the bottom and adds them to the water column (Schelske et al. 1995, Hamilton and Mitchell 1997). The severity of these effects depends upon sediment composition, fetch and presence of macrophyte communities offshore (Schallenberg and Burns 2004). Even at protected beaches, erosion and resuspension may occur from the wakes of boats and personal watercraft; the latter are of particular concern because of their ability to access shallow areas (Asplund 2000). According to the latter author, increased development and visitor use generate increased impacts from motorized watercraft. Yousef et al. (1980) reported significant increases in P and turbidity following motorboat activity.

While MCs are endotoxins, factors which promote cell lysis such as natural senescence or application of algaecides cause MC release into the water (Jones and Orr, 1994). At the same time, cell lysis may increase MC production in the remaining cyanobacterial population (Leao 2012).

The most economical and commonly used algaecide is copper sulphate (Garcia-Villada et al. 2004). Copper concentrations have been found to be inversely correlated with total MC levels (Pip and Allegro 2010). Cyanobacterial sensitivity to copper may be related to catalytic formation of reactive oxygen species, particularly in light (Leunert et al. 2013). The latter workers found that cyanobacteria are more sensitive to copper than green algae that were tested. Thus removal of the dominant cyanobacterial species may be rapidly replaced by blooms of eukaryotic algae (Gibson 1972).

Unfortunately, control achieved with copper sulphate is temporary, and treatments must be repeated. Copper is toxic to and accumulates in a large variety of other aquatic organisms, including macrophytes, zooplankton, molluscs and fish, and is cumulative in the bottom sediments (van Hellebusch et al. 2003, de Oliveira-Filho 2004), which may have long-term adverse effects on the ecosystem (Schuler et al. 2008). Copper sulphate treatment may lead to significant increase in bacterial biomass as a result of plankton death (Havens 1994). Repetitive treatments also promote the growth of copper-resistant mutants in populations of *Microcystis* (Garcia-Villada et al. 2004), and the latter authors have suggested that recurrent use of copper sulphate may eventually lead to its ineffectiveness in controlling cyanobacterial blooms, analogous to the problem of antibiotic resistance in bacteria.

In summary, management of nuisance algal blooms and mitigation of their health risks at beaches include a number of components:

1 Reduction of nutrient sources within the watershed. Nutrients include primarily P, but N may also be a necessary target in N-limited systems.
2 Control of shoreline nutrient sources. These include wastewater and runoff from cottages and recreational areas, and use of fertilizer on lawns and gardens. Adequate sanitary facilities must be provided for beachgoers. Use of ecofriendly personal grooming products should be encouraged. Pets must either be restricted from the public beach, or their waste must be immediately removed. Anglers should discard their bait and clean their fish offsite.
3 Control of shoreline erosion. Measures such as groins, rip-rap and bank stabilization structures reduce the entrainment of nutrients. Conserving and planting vegetation is important in nutrient retention and shoreline stabilization. Designated gravelled or surfaced walkways serve to contain foot traffic and damage from trampled vegetation and dune blow-outs. Use of motorized all-terrain vehicles should be restricted, as well as motorized watercraft in the proximity of beaches.
4 Prompt cleanup of algal accumulations and refuse on the beach. Grooming methods should be chosen such that they do not mix the sand nor incorporate debris below the sand surface.

5 Use of algaecides to control blooms is not recommended as the long term adverse effects on the ecosystem outweigh the temporary relief gained from chemical applications.
6 Public education regarding health hazards posed by blooms and the need to avoid contact or inhalation.
7 Public education regarding beach etiquette and awareness of behaviors which have adverse impacts on the aquatic ecosystem.
8 Monitoring and timely posting of beach closures.
9 Enforcement of beach rules.

References

Auer, M.T., Tomlinson, L.M., Higgins, S.N., Malkin, S.Y., Howell, E.T. and Bootsma, H.A. 2010. Great Lakes *Cladophora* in the 21st century: same algae-different ecosystem. *J. Great Lakes Res.* 36: 248-255.

Backer, L.C., Carmichael, W., Kirkpatrick, B., Williams, C., Irvin, M., Zhou, Y., Johnson, T.B., Nierenberg, K., Hill, V.R., Kieszak, S.M. and Cheng, Y-S. 2008. Recreational exposure at low concentrations of microcystins during an algal bloom in a small lake. *Mar. Drugs* 6: 389-406.

Blaha, L. and Marsalek, B. 2009. Toxins produced in cyanobacterial water blooms – toxicity and risks. *Toxicol.* 2: 36-41.

Blom, G., van Duin, E.H.S. and Lijklema, L. 1994. Sediment resuspension and light conditions in some shallow Dutch lakes. *Wat. Sci. Technol.* 30: 243-252.

Bottcher, G., Chorus, I., Ewald, S., Hintze, T. and Walz, N. 2001. Light-limited growth and microcystin content of *Planktothrix agardhii* and *Microcystis aeruginosa* in turbidostats. In: *Cyanotoxins Occurrence, Causes, Consequences, Chorus*, I., ed. Springer, Berlin. pp. 115-133.

Brown, C.M. and Trick, C.G. 1992. Response of the cyanobacterium, *Oscillatoria tenuis*, to low iron environments: the effect on growth rate and evidence for siderophore production. *Arch. Microbiol.* 157: 349-354.

Byappanahalli, M.N., Shively, D.A., Nevers, M.B., Sadowsky, M.J. and Whitman, R.L. 2006. Growth and survival of *Escherichia coli* and enterococci populations in the macro-alga *Cladophora* (Chlorophyta). *Microbiol. Ecol.* 46: 203-211.

Carmichael, W.W. 1994. The toxins of cyanobacteria. *Sci. Am.* 270: 78-86.

Carmichael, W. 1997. The cyanotoxins. *Adv. Bot. Res.* 27: 211-240.

Carmichael, W.W. and Gorham, P.R. 1981. The mosaic nature of toxic blooms of cyanobacteria. In: *The Water Environment: Algal Toxins and Health*, W.W. Carmichael, ed. Plenum Press, New York. pp. 161-172.

Chorus, I. 2001. Cyanotoxin occurrence in freshwaters – a summary of survey results from different countries. *In Cyanotoxins – Occurrence, causes, consequences*: I. Chorus, ed. Springer-Verlag, Berlin. pp. 75-78.

Chorus, I., Falconer, I., Salas, H. and Bartram, J. 2000. Health risks caused by freshwater cyanobacteria in recreational waters. *J. Toxicol. Environ. Health* B. 3: 323-347.

Correll, D.L. 1998. The role of phosphorus in the eutrophication of receiving waters: a review. *J. Environ. Qual.* 27: 261-266.

Davis, T.W., Berry, D.L., Boyer, G.L. and Gobler, C.J. 2009. The effects of temperature and nutrients on the growth and dynamics of toxic and non-toxic strains of *Microcystis* during cyanobacteria blooms. *Harmful Algae* 8: 715-725.

Dawson, R.M. 1998. The toxicology of microcystins. *Toxicon* 36: 953-962.

de Figueiredo, D.R., Azeiteiro, U.M., Esteves, S.M., Goncalves, F.J.M. and Pereira, M.J. 2004. Microcystin-producing blooms – a serious global public health issue. Ecotoxicol. *Environ. Safety* 59: 151-163.

de Oliveira-Filho, E.C., Lopes, R.M. and Paumgartten, F.J.R. 2004.Comparative study on the susceptibility of freshwater species to copper-based pesticides. *Chemosphere* 56: 369-374.

Dillenberg, H.O. and Dehnel, M.K. 1960. Toxic waterbloom in Saskatchewan, 1959. Can. *Med. Assoc. J.* 83: 1151-1154.

Ding, Y., Song, L. and Sedmak, B. 2013. UVB radiation as a potential selective factor favoring Microcystin producing bloom forming *Cyanobacteria*. PloS One 8(9): e73919.

Downing, J.A., Watson, S.B. and McCauley, E. 2001. Predicting Cyanobacteria dominance in lakes. *Can. J. Fish. Aquat. Sci.* 58: 1905-1908.

Dziallas, C. and Grossart, H-P. 2011. Temperature and biotic factors influence bacterial communities associated with the cyanobacterium *Microcystis* sp. *Environ. Microbiol.* 13:1632-1641.

Englebert, E.T., McDermott, C. and Kleinheinz, G.T. 2008. Effects of the nuisance algae, *Cladophora*, on *Escherichia coli* at recreational beaches in Wisconsin. *Sci. Total. Environ.* 404: 10-17.

Falconer, I.R. 1999. An overview of problems caused by toxic blue-green algae (Cyanobacteria) in drinking and recreational water. *Environ. Toxicol.* 14: 5-12.

Garcia-Villada, L., Rico, M., Altamirano, M., Sanchez-Martin, L., Lopez-Rodas, V. and Costas, E. 2004. Occurrence of copper resistant mutants in the toxic cyanobacteria *Microcystis aeruginosa*: charcterisation and future implications in the use of copper sulphate as algaecide. *Wat. Res.* 38: 2207-2213.

Gibson, C.E. 1972. The algicidal effect of copper on a green and a blue-green alga and some ecological implications. *J.Appl. Ecol.* 9: 513-518.

Gobler, C.J., Davis, T.W., Coyne, K.J. and Boyer, G.L. 2007. The interactive influences of nutrient loading and zooplankton grazing on the growth and toxicity of cyanobacteria blooms in a eutrophic lake. *Harm. Algae* 6: 119-133.

Hamilton, D.P. and Mitchell, S.F. 1997. Wave-induced shear stresses, plant nutrients and chlorophyll in seven shallow lakes. *Freshwat. Biol.* 38: 159-168.

Hansen, P.S., Phlips, E.J. and Aldridge, F.J. 1997. The effects of sediment resuspension on phosphorus available for algal growth in a shallow subtropical lake, Lake Okeechobee. *J. Lake Reserv. Manage.* 13: 154-159.

Havens, K.E. 1994. Structural and functional responses of a freshwater plankton community to acute copper stress. *Environ. Polllut.* 86: 259-266.

Hellstrom, T. 1991. The effects of resuspension on algal production in a shallow lake. *Hydrobiol.* 213: 183-190.

Heuvel, A.V., McDermott, C., Pillsbury, R., Sandrin, T. and Kinzelman, J. 2010. The green alga, *Cladophora,* promotes *Escherichia coli* growth and contamination of recreational waters in Lake Michigan. *Environ. Qual.* 39: 333-344.

Humphries, S.E. and Lyne, V.D. 1988. Blue-green algal blooms: the role of cell buoyancy. *Limnol. Oceanogr.* 33: 79-91.

Izydorczyk, K., Skowron, A., Wojtal, A. and Jurczak, T. 2008. The stream inlet to a shallow bay of a drinking water reservoir, a 'hot-spot' for *Microcystis* blooms initiation. *Int. Rev. Hydrobiol.* 93: 257-268.

Jang, M-H., Ha, K. and Takamura, N. 2008. Microcystin production by *Microcystis aeruginosa* exposed to different stages of herbivorous zooplankton. *Toxicon* 51: 882-889.

Jeppesen, E.M., Sondergaard, M., Meerhoff, M., Lauridsen, T.L. and Jensen, J.P. 2007. Shallow lake restoration by nutrient loading reduction: some recent findings and challenges ahead. *Hydrobiol.* 584: 239-252.

Jepsen, R., Roberts, J. and Lick, W. 1997. Effects of bulk density on sediment erosion rates. Wat. Air Soil Pollut. 99: 21-31.

Jones, G.J. and Orr, P.T. 1994 Release and degradation of microcystin following algicide treatment of a *Microcystis aeruginosa* bloom in a recreational lake, as determined by HPLC and protein phosphatase inhibition assay. *Water Res.* 28: 871-876.

Kearns, K.D. and Hunter, M.D. 2001. Toxin-producing *Anabaena flos-aquae* induces settling of *Chlamydomonas reinhardtii*, a competing motile alga. *Microbiol. Ecol.* 42: 80-86.

Kinzelman, J.L., Whitman, R.L., Byappanahalli, M., Jackson, E. and Bagley, R.C. 2003. Evaluation of beach grooming techniques on *Escherichia coli* density in foreshore sand at North Beach, Racine, WI. *Lake Reserv. Manage*. 19: 349-354.

Lahti, K., Rapala, J., Fardig, M, Niemela, M. and Sivonen, K. 1997. Persistence of cyanobacterial hepatotoxin, microcystin-LR in particulate material and dissolved in lake water. *Water Res.* 31: 1005-1012.

Lampert, W. 1981. Inhibitory and toxic effects of blue-green algae on *Daphnia.* Int. Rev. *Hydrobiol.* 66: 285-298.

Lance, E., Brient, L., Carpentier, A., Acou, A., Marion, L., Bormans, M. and Gerard, C. 2010. Impact of toxic cyanobacteria on gastropods and microcystin accumulation in a eutrophic lake (Grand-Lieu, France) with special reference to *Physa* (=*Physella*) *acuta. Sci. Tot. Environ.* 408: 3560-3568.

Leao, P.N., Engene, N., Antunes, A., Gerwick, W.H. and Vasconcelos, V. 2012. The chemical ecology of cyanobacteria. *Nat. Prod. Rep*. 29: 372-391.

Leunert, F., Grossart, H-P., Gerhardt, V. and Eckert, W. 2013. Toxicant induced changes on delayed fluorescence decay kinetics of Cyanobacteria and green Algae: a rapid and sensitive biotest. *PLoS ONE* 8(4): e63127.

Levich, A.P. 1996. The role of nitrogen-phosphorus ratio in selecting for dominance of phytoplankton by cyanobacteria or green algae and its application to reservoir management. *J. Aquat. Ecosys. Health* 5: 55-61.

Lewis, W.M. and Wurtsbaugh, W.A. 2008. Control of lacustrine phytoplankton by nutrients: erosion of the phosphorus paradigm. Internat. *Rev. Hydrobiol.* 93: 446-465.

Liddle, M.J. and Scorgie, H.R.A. 1980. The effects of recreation on freshwater plants and animals: a review. *Biol. Conserv*. 17: 183-206.

McQueen, D.J. and Lean, D.R.S. 1987. Influence of water temperature and nitrogen to phosphorus ratios on the dominance of blue-green algae in Lake St. George, Ontario. *Can. J. Fish. Aquat. Sci.* 44: 598-604.

Morris, R.J., Williams, D.E., Luu, H.A., Holmes, C.F.B., Andersen, R.J. and Calvert, S.E. 2000. The adsorption of microcystin-LR by natural clay particles. *Toxicon* 38: 303-308.

Murphy, T.P., Irvine, K., Quo, J., Davies, J., Murkin, H., Charlton, M. and Watson, S.B. 2003. New microcystin concerns in the Lower Great Lakes. *Wat. Qual. Res. J. Can.* 38: 127-140.

Neilan, B.A., Pearson, L.A., Muenchhoff, J., Moffitt, M.C. and Dittmann, E. 2013. Environmental conditions that influence toxin biosynthesis in cyanobacteria. *Environ. Microbiol.* 15: 1239-1253.

Ogilvie, B.G. and Mitchell, S.F. 1998. Does sediment resuspension have persistent effects on phytoplankton? Experimental studies in three shallow lakes. *Freshwat. Biol.* 40: 51-63.

Oh, H.M., Lee, S.J., Kim, J.H., Kim, H.S. and Yoon, B.D. 2001. Seasonal variation and indirect monitoring of microcystin concentrations in Daechung reservoir, Korea. *Appl. Environ. Microbiol.* 67: 1484-1489.

Oh, H.M., Lee, S.J., Jang, M.H. and Yoon, B.D. 2000. Microcystin production by *Microcystis aeruginosa* in a phosphorus-limited chemostat. *Appl. Environ. Microbiol.* 66: 176-179.

Olapade, O.A., Depas, M.M., Jensen, E.T. and McLellan, S.L. 2006. Microbial communities and fecal indicator bacteria associated with *Cladiophora* mats on beach sites along Lake Michigan shores. *Appl. Environ. Microbiol.* 72: 1932-1938.

Orihel, D.M., Bird, D.F., Brylinsky, M., Chen, H., Donald, D.B., Huang, D.Y., Giani, A., Kinniburgh, D., Kling, H., Kotak, B.G., Leavitt, P.R., Nielsen, C.C., Reedyk, S., Rooney, R.C., Watson, S.B., Zurawell, R.W. and Vinebrooke, R.D. 2012. High microcystin concentrations occur only at low nitrogen-to-phosphorus ratios in nutrient-rich Canadian lakes. *Can. J. Fish. Aquat. Sci.* 69: 1457-1462.

Orr, P.T. and G.J. Jones. 1998. Relationship between microcystin production and cell division rates in nitrogen-limited *Microcystis aeruginosa* cultures. *Limnol. Oceanogr.* 43: 1604-1614.

Paerl, H.W. 1988. Nuisance phytoplankton blooms in coastal, estuarine and inland waters. *Limnol. Oceanogr.* 33: 823-847.

Paerl, H.W. and Otten, T.G. 2013. Blooms bite the hand that feeds them. *Science* 342: 433-434.

Pawlik-Skowronska, B., Kalinowska, R. and Skowronski, T. 2013. Cyanotoxin diversity and food web bioaccumulation in a reservoir with decreasing phosphorus concentrations and perennial cyanobacterial blooms. *Harm. Algae* 28: 118-125.

Pick, F.R. and Lean D.R.S. 1987. The role of macronutrients (C,N,P) in controlling cyanobacterial dominance in temperate lakes. *N. Zeal. J. Mar. Freshwat. Res.* 21: 425- 434.

Pip, E. 2005. Surface water quality in Manitoba with respect to six chemical parameters, water body and sediment type and land use. *Aquat. Ecosys. Health Manage.* 8: 195- 207.

Pip, E. 1988. Chlorophyll concentrations in relation to environmental parameters in different situations in Shoal Lake (Manitoba-Ontario). *Int. Revue ges. Hydrobiol.* 73: 417-429.

Pip, E. and Allegro, E. 2010. Nearshore fluctuations in water chemistry, microcystins and coliform bacteria during the ice-free season in Lake Winnipeg, Manitoba, Canada. *Ecohydrol. Hydrobiol.* 10: 35-43.

Pip, E. and Bowman, L. submitted. *Microcystin and algal chlorophyll in relation to nearshore nutrient concentrations in Lake* Winnipeg, Manitoba, Canada

Rapala, J., Sivonen, K., Lyra, C. and Niemela, S.I. 1997. Variation of microcystins, cyanobacterial hepatotoxins, in *Anabaena* spp. as a function of growth stimuli. *Appl. Environ. Microbiol.* 63: 2206-2212.

Rapala, J. and Sivonen, K. 1998. Assessment of environmental conditions that favor hepatotoxic and neurotoxic *Anabaena* spp. strains cultured under light limitation at different temperatures. *Microb. Ecol.* 36: 181-192.

Rastogi, R.P. and Sinha, R.P. 2009. Biotechnological and industrial significance of cyanobacterial secondary metabolites. *Biotechnol. Adv.* 27: 521-539.

Rinta-Kanto, J.M., Ouellette, A.J.A., Twis, M.R., Boyer, G.L., Bridgeman, T. and Wilhelm, S.W. 2005. Quantification of toxic *Microcystis* spp. during the 2003 and 2004 blooms in western Lake Erie using quantitative real time PCR. *Environ. Sci. Technol.* 39: 4198-4205.

Schallenberg, M. and Burns, C.W. 2004. Effects of sediment resuspension on phytoplankton production: teasing apart the influences of light, nutrients and algal entrainment. *Freshwat. Biol.* 49: 143-159.

Schelske, C.L., Carrick, H.J. and Aldridge, F.J. 1995. Can wind-induced resuspension of meroplankton affect phytoplankton dynamics? *J. North Amer. Benthol. Soc.* 14: 616-630.

Schindler, D.W., Hecky, R.E., Findlay, D.L., Stainton, M.P., Parker, B.R., Paterson, M.J., Beaty, K.G., Lyng, M. and Kasian, S.E.M. 2008. Eutrophication of lakes cannot be controlled by reducing nitrogen input: Results of a 37-year whole-ecosystem experiment. *Proc. Nat. Acad. Sci.* 105: 11254-11258.

Schuler, L.J., Hoang, T.C. and Rand, G.M. 2008. Aquatic risk assessment of copper in freshwater and saltwater ecosystems of South Florida. *Ecotoxicol*. 17: 642-659.

Sinang, S.C., Reichwaldt, E.S. and Ghadouani, A. 2013. Spatial and temporal variability in the relationship between cyanobacterial biomass and microcystins. *Environ. Monit Assess*. 185: 6379-6395.

Sivonen, K. 1990: Effects of light, temperature, nitrate, orthophosphate and bacteria on growth of and hepatotoxin production by *Oscillatoria agardhii* strains. *Appl. Environ. Microbiol*. 56: 2658-2666.

Smith, V.H. 1983. Low nitrogen to phosphorus ratios favor dominance by blue-green algae in lake phytoplankton. *Science* 221: 669-671.

Sonzogni, W.C., Chapra, S.C., Armstrong, D.E. and Logan, T.J. 1982. Bioavailability of phosphorus inputs to lakes. *J. Environ. Qual*. 11: 555-563.

Stanley, C.D., Clarke, R.A., McNeal, B.L. and MacLeod, B.W. 2003. *Relationship of chlorophyll a concentration to seasonal water quality in Lake Manatee, Florida*. Soil and Water Science Department, Florida Cooperative Extension Service, University of Florida. Publ. SL-211.

Stewart, L., Webb, P.M., Schluter, P.J. and Shaw, G.R. 2006. Recreational and occupational field exposure to freshwater cyanobacteria – a review of anecdotal and case reports, epidemiologic studies, and the challenges for epidemiologic assessment. *Environ. Health* 5: 6.

Te, S.H. and Gin, K.Y-H. 2011. The dynamics of cyanobacteria and microcystin production in a tropical reservoir of Singapore. *Harmful Algae* 10: 319-329.

Tsuji, K., Nalto, S., Kondo, F., Ishikawa, N, Watanabe, M.F., Suzuki, M. and Harada, K. 1994. Stability of microcystins from Cyanobacteria: effect of light on decomposition anf isomerization. *Environ. Sci. Technol*. 28: 173-177.

Utkilen, H. and Gjolme, N. 1992. Toxin production by *Microcystis aeruginosa* as a function of light in continuous cultures and its ecological significance. *Appl. Environ. Microbiol*. 58: 1321-1325.

Van Hellebusch, E., Chatenet, P., Deluchar, V., Chazal, P.M., Froissard, D., Botineau, M., Ghestem, A. and Baudu, M. 2003. Copper accumulation in a reservoir ecosystem following copper sulfate treatment (St. Germain Les Belles, France). *Wat. Air Soil Pollut*. 150: 3-22.

Vazquez-Coriano, L. 2011. *Developing a tool to help states address cyanobacteria harmful algal blooms (HABs) in freshwater and drinking water*. 2010-2011. Environmental Public Health Leadership Institute, Washington DC.

Vezie, C., Rapala, J., Vaitomaa, J., Seitsonen, J., and Sivonen, K. 2002. Effect of nitrogen and phosphorus on growth of toxic and nontoxic *Microcystis* strains and on intracellular microcystin concentrations. *Microb. Ecol.* 43: 443-454.

Watanabe, M.F. and Oishi, S. 1985. Effects of environmental factors on toxicity of a cyanobacterium (*Microcystis aeruginosa*) under culture conditions. *Appl. Environ. Microbiol.* 49: 1342-1344.

Watson, S.B. 2003. Cyanobacterial and eukaryotic algal odour compounds: signals or by- products? A review of their biological activity. *Phycol.* 42: 332-350.

Watson, S.B., Ridal, J. and Boyer, G.L. 2008. Taste and odour and cyanobacterial toxins: impairment, prediction, and management in the Great Lakes. *Can. J. Fish. Aquat. Sci.* 65: 1779-1796.

Whitman, R.L., Shively, D.A., Pawlik, H., Nevers, M.B. and M.N. Byappanahalli. 2003. Occurrence of *Escherichia coli* and enterococci in *Cladophora* (Chlorophyta) in nearshore water and beach sand of Lake Michigan. *Appl. Environ. Microbiol.* 69: 4714-4719.

Wicks, R.J. and Thiel, P.G. 1990. Environmental factors affecting the production of peptide toxins in floating scums of the cyanobacterium *Microcystis aeruginosa* in a hypertrophic African reservoir. *Environ. Sci. Technol.* 24: 1413-1418.

Wiegand, C. and Pflugmacher, S. 2005. Ecotoxicological effects of selected cyanobacterial secondary metabolites: a short review. *Toxicol. Appl. Pharmacol.* 203: 201-218.

Wormer, L., Huerta-Fontela, M., Cires, S., Carrasco, D. and Quesada, A. 2010. Natural photodegradation of the cyanobacterial toxins microcystin and cylindrospermopsin. *Environ. Sci. Technol.* 44: 3002-3007.

Yousef, Y.A., McLellon, W.M. and H.H. Zebuth. 1980. Changes in phosphorus concentrations due to mixing by motor boats in shallow lakes. *Wat. Res.* 14: 841-852.

Yu, S. 1995. Primary prevention of hepatocellular carcinoma. *J. Gastroenterol. Hepatol.* 10: 674-682.

Zurawell, R.W., Chen, H., Burke, J.M. and Prepas, E.E. 2005. Hepatotoxic Cyanobacteria: A review of the biological importance of microcystins in freshwater environments. J. *Toxicol. Environ. Health, Part B: Critical Reviews* 8: 1-37.

In: Beaches
Editor: Margaret Cessa

ISBN: 978-1-63117-239-7

Chapter 6

DOWNCOAST EROSION TRIGGERED BY EXHAUSTION OF SAND SUPPLY FROM SEA CLIFFS: AN EXAMPLE OF NORTHERN KUJUKURI BEACH IN JAPAN

Takaaki Uda
Public Works Research Center, Taito, Tokyo, Japan

The relationship between the protection of sea cliffs from which a large amount of sand was supplied to the downcoast and the resultant downcoast erosion was investigated, taking the northern Kujukuri Beach as an example. A seawall was constructed along with the installation of concrete blocks to protect the downcoast locally, resulting in shoreline recession further downcoast. These actions were recurrently carried out on this coast, and a long stretch of the coastline was covered with concrete blocks, causing the loss of a natural sandy beach. These beach changes were reproduced using the BG model (a three-dimensional model for predicting beach changes based on Bagnold's concept). On the basis of these field data, future measures were considered.

INTRODUCTION

Beach erosion caused by anthropogenic factors has been severe at many coasts in Japan. Uda (2010) has classified the cause of beach erosion into seven categories. One of them is in the selection of the measure itself, i.e, the protection of sea cliffs from which a large amount of sand was supplied to the downcoast, resulting in a marked decrease in sand supply to the downcoast and subsequent beach erosion. When downcoast erosion occured, a seawall was constructed along with the installation of concrete blocks to protect the site locally, resulting in shoreline recession further downcoast. These actions were recurrently carried out, and a long stretch of the coastline was covered with concrete blocks, causing the loss of a natural sandy beach. In this study, this issue is investigated, taking northern Kujukuri Beach, as shown in Fig. 1, as an example.

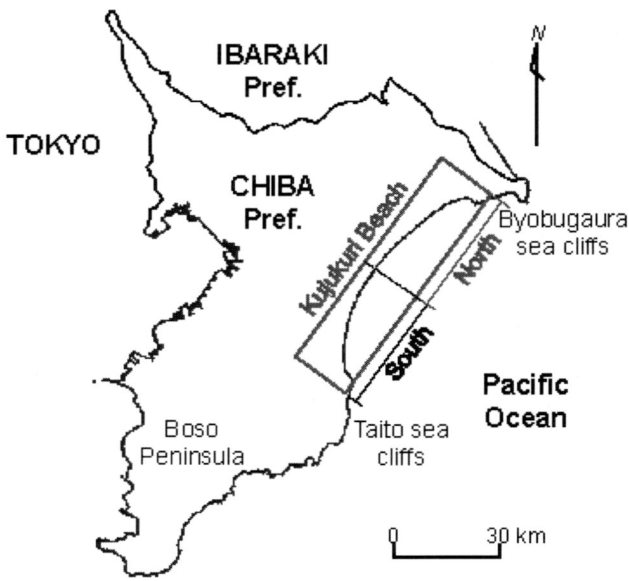

Figure 1. Location of Kujukuri Beach facing Pacific Ocean.

This beach is a 60-km-long sandy beach facing the Pacific Ocean and extending from Byobugaura and Taito sea cliffs located at the north and south ends, respectively, and has been formed by the deposition of sand supplied from the sea cliffs mainly composed of unconsolidated mud and sand stones (Horikawa and Sunamura, 1971). The Byobugaura sea cliffs, which constitute the main source of sand, had been retreating at a rate of 0.7 m/yr and much sediment was supplied to the coast (Horikawa and Sunamura, 1969; 1970). However, sand supply to Kujukuri Beach markedly decreased mainly because of the installation of concrete blocks along the sea cliffs as a measure against erosion since the 1960s. Longshore sand transport was also obstructed by the construction of Iioka, Taito, and Katagai fishing ports located at the north and south ends of the beach, and at the central part of the beach, respectively. Moreover, the volume of littoral sand was decreased by the offshore disposal of dredged materials at these fishing ports. All these activities resulted in the expansion of the eroded zone from the vicinity of both ends of the pocket beach to the central part. Thus, this beach is in critical condition at present in terms of the preservation of the natural sandy beach. Here, measured beach changes were reproduced using the BG model (a three-dimensional model for predicting beach changes based on Bagnold's concept) proposed by Serizawa et al. (2007), and future measures were considered on the basis of the results of the numerical simulation.

SHORELINE CHANGES

Figure 2 shows an aerial photograph of the northern Kujukuri Beach in 2010 extending north of Katagai fishing port, as shown in Fig. 1. A gradually curved embayment shoreline extends from Iioka fishing port located at the north end of the study area to Katagai fishing port in the northeast-southwest direction. The shoreline changes in the entire area, as shown in Fig. 2, were investigated using aerial photographs. Figure 3 shows the shoreline changes between 1947 and 2010, where the shoreline changes were read to the direction normal to the shoreline in 2010 between Iioka fishing port and the Shimonagai coast because the shoreline has a large curvature, as shown in Fig. 2.

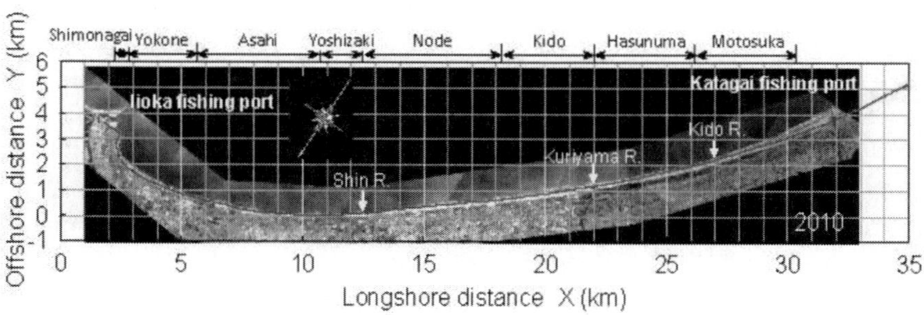

Figure 2. Aerial photograph of northern Kujukuri Beach between Iioka and Katagai fishing ports in 2010.

Figure 3 demonstrates that the area with a wide foreshore and that with no sandy beach are in remarkable contrast with each other, and the area where the shoreline advanced concentrated on the Shimonagai coast and in the area northeast of Katagai fishing port with maximum shoreline advances of 226 and 385 m at the Shimonagai coast and Katagai fishing port, respectively. Except for the local shoreline advance at the Yashigaura coast (X = 6.5 km) and the Kuriyama River mouth (X = 22 km), almost the entire area between two sand accumulation zones is erosive, the sandy beach completely disappeared between X = 4.3 and 14.8 km, and the seawall was exposed to waves. In the area where the sandy beach is left, erosion on the Node coast was most severe, and a maximum shoreline recession of 56 m occurred at a location of X = 15.2 km.

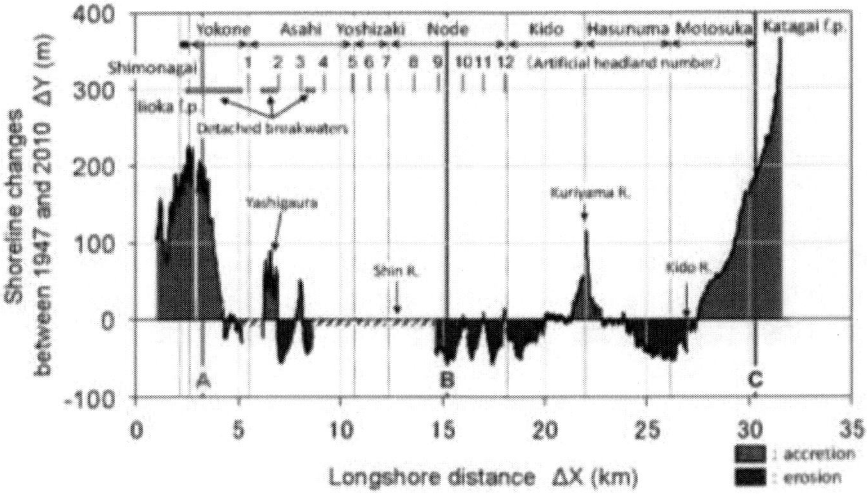

Figure 3. Shoreline changes between 1947 and 2010.

In 1947, Kujukuri Beach was a natural sandy beach in which sand supplied from Byobugaura northeast of the beach was transported with no obstruction by artificial structures toward the central part by longshore sand transport. Then, the Iioka fishing port breakwater was extended, as shown in Fig. 4, and the south breakwater of the fishing port was extended 1 km offshore from the previous coastline until 2010. Owing to the extension of the south breakwater, westward longshore sand transport was obstructed, and a large wave-shelter zone was formed west of the south breakwater, causing sand deposition in the wave-shelter zone, and severe erosion occurred downcoast because of marked decrease in sand supply from the upcoast.

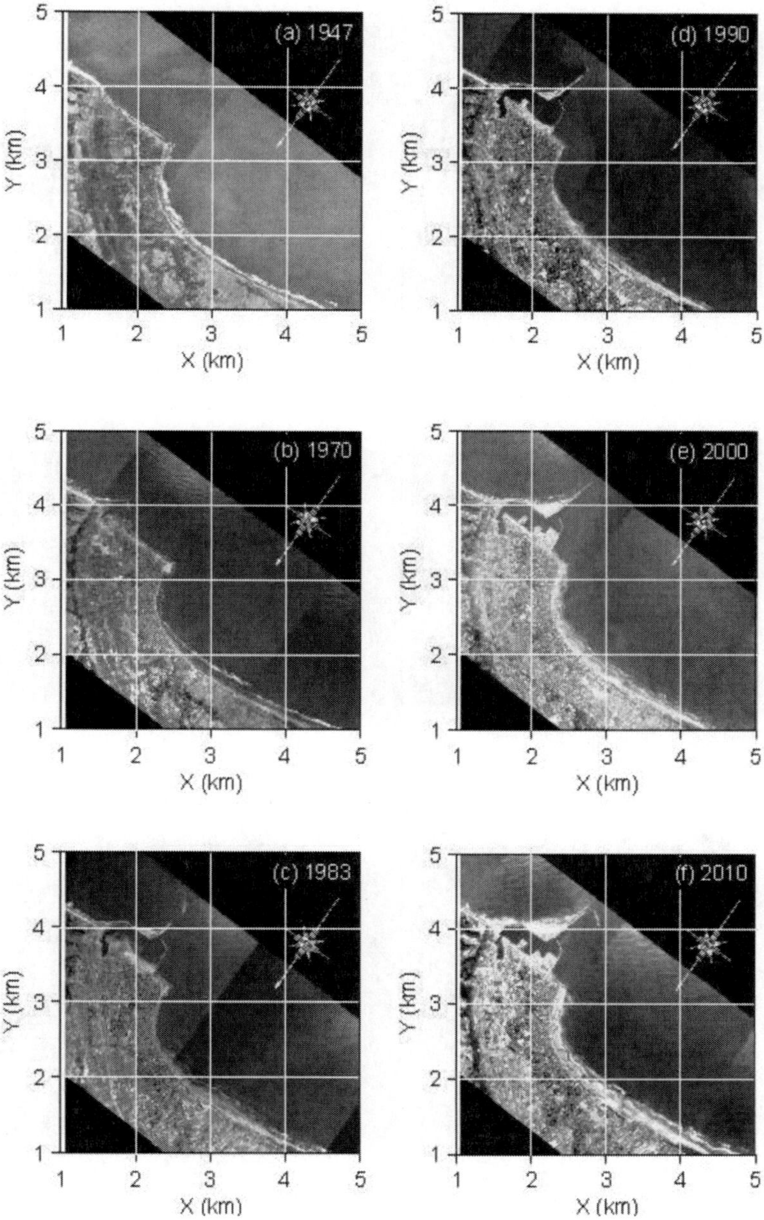

Figure 4. Aerial photographs of Iioka fishing port and nearby coast.

On further downcoast near Katagai fishing port, a large amount of sand was deposited between the Motosuka coast and Katagai fishing port, as shown in Fig. 3, because of the obstruction of longshore sand transport by the Katagai fishing port breakwater, which was extended over time, as shown in Fig. 5.

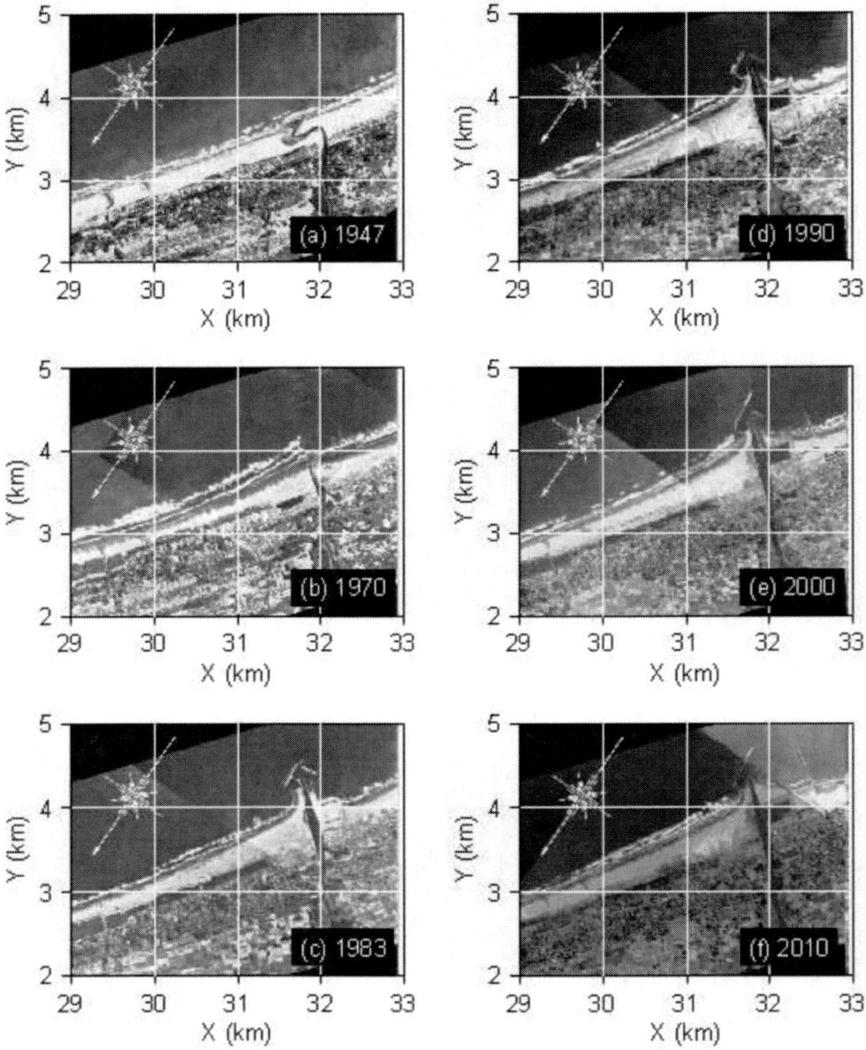

Figure 5. Aerial photographs of Katagai fishing port and nearby coast.

Figure 6 shows the change in shoreline position between 1947 and 2010 at points A, B and C on the Yokone, Node, and Motosuka coasts, respectively, where marked shoreline changes were observed in Fig. 3. Although no shoreline changes had been observed before 1970, the shoreline at point A advanced at a rate of 6.8 m/yr with the construction of the Iioka fishing port breakwater, and 15 detached breakwaters on the Iioka coast, resulting in the total shoreline advance of 205 m until 2010. In contrast, the shoreline position at point B retreated by a rate of 1.3 m/yr with a 13 year time lag compared with that at point A with the total shoreline recession of 57 m until 2010. In addition, the shoreline at point C advanced by a rate of 5.8 m/yr with the total shoreline advance of 176 m until 2010. At all points, the shoreline still continues to change at present.

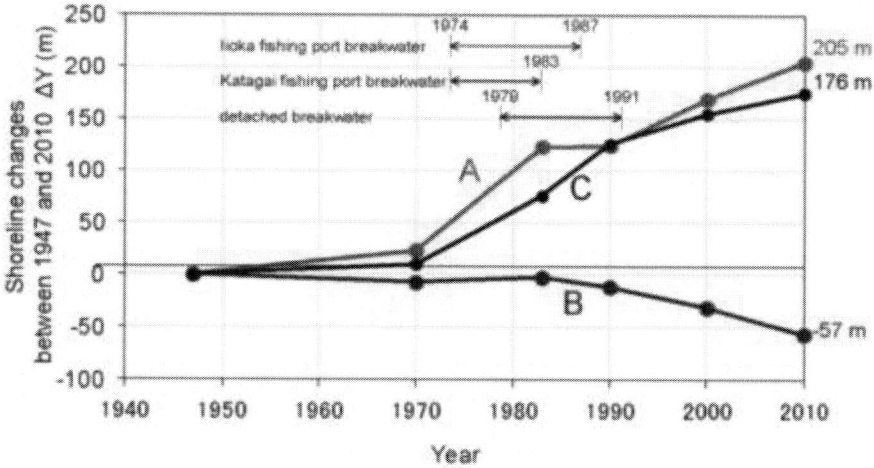

Figure 6. Shoreline changes with time at location A of Iioka, location B of Node and location C of Motosuka coasts.

Sand supplied from Byobugaura was transported westward under natural conditions, and the longshore sand transport was estimated to be 3.0×10^5 m^3/yr between 1960 and 1967 (Uda, 1997). Then, the breakwater of Iioka fishing port was extended, resulting in the reduction in sand supply to the west of the fishing port, and sand supply from the sea cliffs was assumed to be reduced to 9×10^4 m^3/yr since 1967 (Uda, 1997). Furthermore, Uda et al. (2010) estimated that the longshore sand transport rate obstructed by the south breakwater was 1.4×10^4 m^3/yr on average between 1985 and 2005 after the extension of the south breakwater, and the same amount of longshore sand transport discharged downcoast, while turning around the tip of the breakwater after 2005. Note that the rate of longshore sand transport is one order of magnitude smaller than that under the natural conditions in the past.

At Katagai fishing port, a 400-m-long jetty was extended in 1970 at the Sakuta River mouth, and then in 1983, two oblique breakwaters were constructed at the tip of the jetty to prevent sand deposition, as shown in Fig. 5. Thereafter, the Katagai fishing port breakwater was further extended until 2000, resulting in further shoreline advance northeast of the port because of the blockage of longshore sand transport. Regarding the beach changes around Katagai fishing port, San-nami et al. (2009) estimated longshore sand transport turning around the tip of the breakwater to be 7×10^4 m^3/yr on the basis of a numerical simulation using the contour-line-change model proposed by Serizawa et al. (2003). Thus, as the sand budget in the study area, sand is supplied from upcoast (Iioka fishing port) at a rate of 1.4×10^4 m^3/yr, whereas sand is transported away from the downcoast boundary (Katagai fishing port) at a rate of 7×10^4 m^3/yr with no balance in sand transport, implying that erosion became severe with time.

CHANGE IN BEACH WIDTH

On Kujukuri Beach the seaward advance of the marginal line of the coastal forest significantly affected to the decrease in the beach width (Hoshigami et al., 2005). Therefore, the shoreline position and location of the seaward marginal line were read from the aerial photographs taken between 1947 and 2010, and the change in beach width was calculated (Fig. 7). The vertical axis shows the offshore distance from the location of the seawall to the shoreline and the marginal line of the coastal forest in 2010. From the definition of the variables, the distance between the shoreline and the marginal line of the coastal forest corresponds to the beach width.

Although the beach width was as narrow as 28 m in the vicinity of Iioka fishing port in 1947, when a natural sandy beach extended, it gradually expanded southwestward with a maximum beach width of 175 m on the Motosuka coast at X = 26. 4 km. By 1970, the breakwaters of Iioka and Katagai fishing ports had been constructed, and the shoreline started to advance northeast of the breakwater. However, as a result of the seaward advance of the marginal line of the coastal forest, which occurred simultaneously with the construction of the breakwaters, the beach width was narrowed.

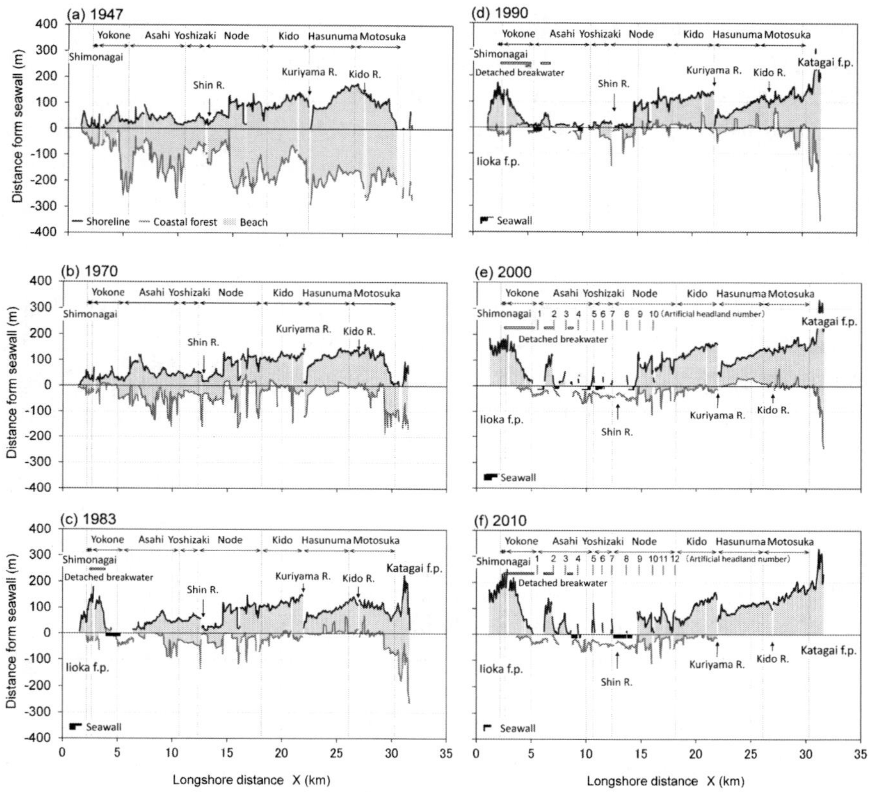

Figure 7. Offshore distance to shoreline and marginal line of coastal forest from seawall.

By 1983, the Iioka fishing port breakwater was extended along with the construction of seven detached breakwaters on the Yokone coast as a measure against beach erosion, and then the beach width behind these detached breakwaters extended to a maximum of 192 m at a location of $X = 2.8$ km. In the area between the Yokone and Asahi coasts ($X = 4.5$ and 6.4 km) further downcoast of the detached breakwaters, the sandy beach was eroded away and the seawall was exposed to waves. Moreover, the Katagai fishing port breakwater was further extended, resulting in further sand deposition and shoreline advance upcoast of the fishing port owing to the obstruction of longshore sand transport.

By 1990 and 2000, the south breakwater of Iioka fishing port was further extended along with the construction of 12 detached breakwaters and 10 artificial headlands. The beach was further widened between the Shimonagai and Yokone coasts (X =2.2 and 5.3 km), whereas on the Asahi and Node coasts, the sandy beach disappeared. The area where the sandy beach disappeared was between X = 4.7 and 10.6 km (extension 5.9 km) in 1990; it expanded to the area between X = 5.5 and 14.3 km (extension 8.8 km) until 2000, resulting in the southwestward expansion of the area with the exposed seawall. Furthermore, the additional extension of the Katagai fishing port breakwater caused the shoreline advance on the northeast side of the fishing port.

By 2010, the number of detached breakwaters increased to 15 between the Shimonagai and Asahi coasts along with the construction of 12 artificial headlands further downcoast. Between the Shimonagai and Yokone coasts (X = 2.2-5.3 km), the sandy beach width further extended, and simultaneously, the sandy beach completely disappeared between the Asahi and Node coasts (X = 5.5-14.6 km), resulting in the exposure of the seawall along 9.1 km length, which is equal to one-third of all the northern Kujukuri Beach. The length where the seawall was exposed to waves gradually expanded downcoast between 1983 and 2010 with a mean expansion velocity of 270 m/yr.

The reduction of the beach width between the shoreline and the marginal line of the coastal forest was triggered by both the shoreline recession (landward movement) and the advance of the marginal line of the coastal forest (seaward movement). Of these two factors related to the reduction in beach width, the plantation of coastal forest had been extensively carried after WWII to stabilize the devastated coastal land and to reduce windblown sand. However, the excess advance in the coastal forest zone up to the vicinity of the shoreline became one of the causes of the reduction in sandy beach width as a buffer zone against beach changes due to waves (Uda, 2010).

SITE OBSERVATION

To investigate the present condition of the northern Kujukuri Beach, site observation was carried out on November 18 and 19, 2011. Figure 8 shows a photograph of Byobugaura, which is the main supply source of sand to Kujukuri Beach, taken from the shoreline east of Iioka fishing port. Although Byobugaura is made up of the sea cliffs with 30-50 m height and composed of unconsolidated layers of sand and mud stones, a number of concrete blocks have been constructed continuously along the sea cliffs to prevent erosion. This induced a marked reduction in sand supply to the downcoast, resulting in beach erosion.

Figure 9 shows a photograph of the long south breakwater of Iioka fishing port taken from Gyobu Point located at the southwest end of Byobugaura. In the past, with no breakwaters of the fishing port, westward longshore sand transport freely passed this location, but now, it is obstructed by the breakwater. The formation of the sandy beach on the left (east) side of the breakwater is evidence of the blocking of smooth longshore sand transport.

Figure 8. Photograph of Byobugaura Sea Cliffs taken from shoreline east of Iioka fishing port.

Figure 9. Long breakwater of Iioka fishing port, looking west from Gyobu Point located at southwest end of Byobugaura.

Figure 10. Seawall undermined by discharge of fill material behind seawall at a location of $X = 9$ km on Asahi coast.

Figure 11. Seawall and a number of concrete blocks ($X = 13$ km, Node coast, looking north).

Although the foreshore completely disappeared between $X = 8.6$ and 14.6 km in Fig. 3 and the seawall was completely exposed to waves, the seawall was undermined at $X = 9$ km owing to the discharge of the fill material behind the seawall, which barely collapsed, as shown in Fig. 10. The concrete blocks in front of the seawall also subsided and their wave dissipation effect was lost. Similarly, at a location of $X = 13$ km on the Node coast, where the beach was severely eroded, a number of large concrete blocks were piled to the level by which ocean view from the shoreline was totally lost, as shown in Fig. 11. Further south, the seawall and their foot protection extended straight for a long distance (Fig. 12).

Figure 12. Seawall and subsided concrete blocks (X = 13 km, Node coast, looking south).

Figure 13. Scarp of 1.5 m height formed downcoast of artificial headland No. 9 on Node coast.

Figure 14. Isolated dune vegetation of an agave left on backshore due to erosion.

Sandy beach reappeared southwest of artificial headland No. 9 on the Node coast, but erosion was under way at that site. Immediately south of artificial headland No. 9, a scarp of 1.5 m height was formed, as shown in Fig. 13, and downcoast of this area, an isolated dune vegetation of an agave, which originally grew on the sand dune, was left on the backshore owing to erosion, as shown in Fig. 14.

Figure 15. Beach erosion in front of playground constructed south of artificial headland No. 11.

Figure 16. Collapsed gently sloping revetment (X = 19 km, Kido coast).

Figure 17. Natural sandy beach with a very gentle slope of 1/50 left south of the Kido River mouth.

At a location 600-850 m downcoast of artificial headland No. 11 on the Node coast, there was a wide sandy beach in the past, and therefore, a playground was constructed very close to the shoreline along with the construction of the seawall to protect the facility. However, this seawall was exposed to waves by successive beach erosion, which in turn caused the obstruction of southward longshore sand transport. As a result, a scarp of 1.8 m height was formed immediately downcoast of the seawall (Fig. 15).

Figure 16 shows the damaged gently sloping revetment built at a location of $X = 19$ km on the Kido coast. Almost entire concrete blocks on the slope of the revetment subsided and the revetment was severely destroyed. Because the original elevation of the crown concrete blocks was 0.2 m below the top of the revetment, where a white concrete wall can be seen in Fig. 16, it is clear that the concrete blocks subsided by 0.7 m. The collapse of the gently sloping revetment is mainly caused by the subsidence of the foot of the revetment associated with the erosion and the resultant discharge of the fill material under the revetment (Uda et al., 2005). It is clear that the gently sloping revetment is easily destroyed when the foot of the revetment is exposed to waves after the shoreline recession. However, sandy beach with a very gentle foreshore slope of 1/50, being the lost scene of the past Kujukuri Beach, was left south of the Kido River mouth located at $X = 27$ km in contrast to the severe erosion north of there, as shown in Fig. 17.

NUMERICAL SIMULATION USING BG MODEL

Method and Calculation Conditions

The BG model proposed by Serizawa et al. (2007) was employed to predict beach changes. The calculation domain was a 33-km-long coastline extending between Iioka and Katagai fishing ports, as shown in Fig. 18. As the boundary conditions, longshore sand transport of 1.4×10^4 m^3/yr, which was estimated by Uda et al. (2010), was assumed at the northeast boundary ($X = 0$ km), whereas a free boundary was set at the southwest boundary.

For the wave conditions, the energy-mean significant wave height measured offshore of Choshi was adopted and seasonal variations in wave characteristics were considered; relatively calm waves with $H = 1.5$ m ($T = 11.3$ s) and rough waves with $H = 2.0$ ($T = 7.9$ s) were considered with the predominance of southerly waves from the direction of N113° E and northerly waves from the direction of N79° E in summer and winter, respectively.

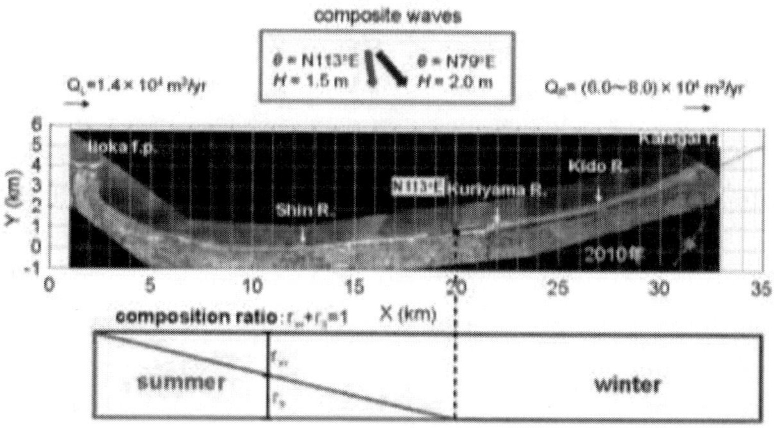

Figure 18. Composite waves for calculation.

The coastline of north Kujukuri Beach between Iioka and Katagai fishing ports has a gradually curved concave shape, as shown in Fig. 18, and the wave direction coincides with the direction normal to the shoreline, N113° E, in the area between X = 20 and 32 km in summer, resulting in the nil longshore sand transport. In contrast, the angle between the direction normal to the shoreline and wave direction becomes large in the area between X = 2 and 20 km, resulting in large longshore sand transport. In other words, the area where summer waves markedly affect the shoreline change becomes the shoreline between X = 2 and 20 km. Under the condition of the winter wave direction of N79° E, a large wave-shelter zone is formed west of Iioka fishing port, and longshore sand transport reduces to zero because of a significant decrease in wave height in the wave-shelter zone. This implies that the effect of winter waves on the beach changes is weak in the wave-shelter zone of Iioka fishing port, but its effect increases with increasing southwestward distance from the fishing port. Taking the effect of the seasonal change in wave conditions on the shoreline changes into account, composite waves composed of summer and winter waves were assumed in the study area, so that the effects of summer and winter waves are predominant in the area between Iioka fishing port and X = 20 km, and in the area between X = 20 km and Katagai fishing port, respectively. First, longshore distribution of wave height and wave direction was calculated using the angular spreading method for irregular waves (Sakai et al., 2006) given incident waves in summer and winter. Then, the composite waves in summer and winter were considered; the weight of summer waves (r_s) takes 1.0 at Iioka fishing port, linearly reduces southwestward, and takes 0 at X = 20 km. The weight function (r_w) in winter is set in the opposite way, as shown in Fig. 18. The composite waves were obtained on the basis of wave height and wave direction at any location multiplied by the weight functions of r_w and r_s, as shown in Fig. 18.

The depth of closure h_c was determined using Eq. (1) assuming that a linear relationship stands between h_c and wave height H at a local point.

$$h_c = aH \tag{1}$$

Here a is a constant. Because the depth of closure is approximately given as 10 m around Katagai fishing port owing to the bathymetries near the fishing port, and the energy-mean wave height in winter, which determines offshore beach changes, is H_{max} = 2 m, constant a becomes 5.0. Figure 19 shows the distribution of h_c. Although h_c is equal to 10 m in almost the entire area, it decreases in the wave-shelter zone of Iioka fishing port.

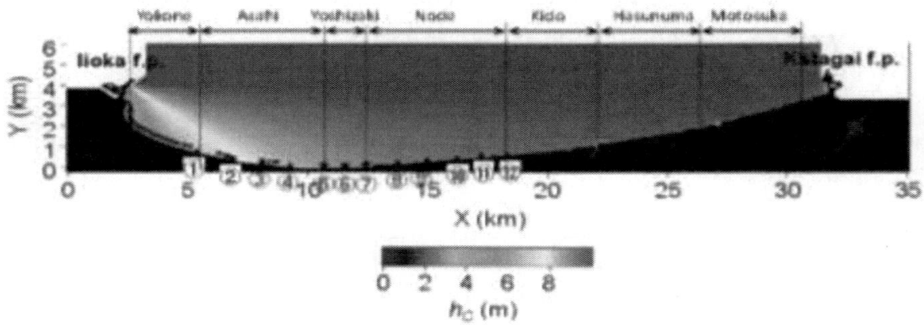

Figure 19. Planar distribution of depth of closure h_c.

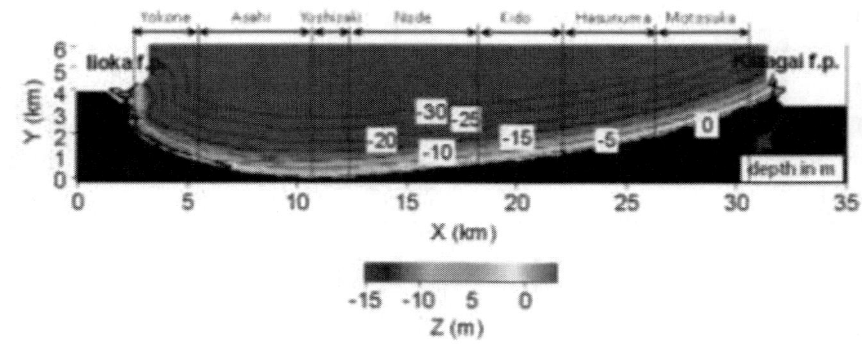

Figure 20. Initial beach topography.

As the initial beach topography, the topography in 1983 was given, when the foot protection works against cliff erosion were completed and the Iioka fishing port breakwater was extended. Because the bathymetry in the entire north Kujukuri Beach was not measured in 1983, the bathymetry in 1983 was assumed on the basis of the shoreline configuration in 1983 given the mean beach slope of 1/60. In the reproduction calculation, the construction history of the coastal and port structures in 1990, 2000 and 2010 was considered, as shown in Fig. 21. Table 1 shows the calculation conditions.

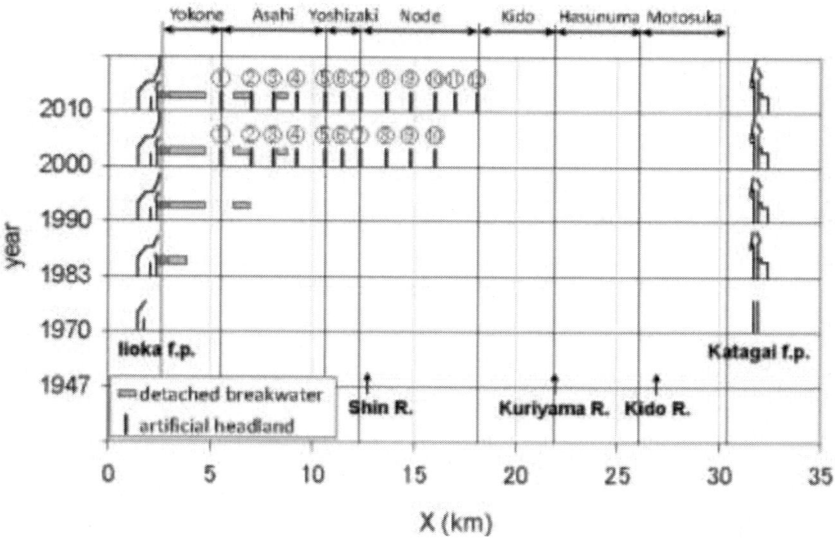

Figure 21. Construction history of coastal structures.

Table 1. Calculation conditions

Calculation method	BG model (Serizawa et al., 2007)
Study area	North Kujukuri Beach
Calculation cases	Reproduction of beach topographies in 1990, 2000 and 2010
Initial topography	Produced bathymetry on the basis of shoreline configuration in 1983 given the mean beach slope of 1/60
Equilibrium slope	1/60
Sea level	Mean sea level
Wave conditions	Compound waves of summer waves with a breaker height H_b = 1.5 m and α_b = 32° (N113° E) and winter waves with a breaker height H_b = 2.0 and α_b = 63° (N79° E)

Table 1. (Continued)

Depth of closure and berm height	$h_c = 5.0H$ and $h_R = 3$ m
Boundary conditions	1.0×10^4 m^3/yr at left boundary and free boundary at right boundary
Coefficients of sand transport	Coefficient of longshore sand transport $A = 0.5$ Ozasa and Brampton's coefficient $\varsigma = 1.62$ Coefficient of cross-shore sand transport to longshore sand transport $K_z/K_x = 0.7$
Depth distribution of sand transport	Qubic equation of depth as in Serizawa et al. (2007)
Critical slope of falling sand	1/2 on land, 1/3 under the sea surface
Calculation domain	35 km length in longshore direction and 6 km in cross-shore direction
Mesh size	$\Delta X = 100$ m and $\Delta Y = 10$ m
Time intervals	$\Delta t = 1$ hr
Calculation method	Explicit finite difference method

Calculation Results

The reproduced bathymetry in 1990 and the bathymetric changes with reference to that in 1983 are shown in Fig. 22(a) togerther with longshore sand transport at several points. By 1990, the seabed depth had increased in the zone shallower than -10 m in the area between $X = 7$ and 28 km, whereas sand was deposited on the Yokone coast west of Iioka fishing port and on the Motosuka coast near Katagai fishing port. By 2000, erosion and accretion were further intensified compared with those in 1990 (Fig. 22(b)). Although artificial headlands (HL) Nos. 1-10 had been constructed by 1990 as a measure against beach erosion, the decrease in seabed elevation still continued in the zone shallower than -10 m on the Asahi and Yoshizaki coasts. By 2010, erosion was further intensified and the erosion zone markedly expanded westward, resulting in the decrease in seabed elevation between the Asahi coast and the central part of the Kido coast (Fig. 22(c)). In contrast, sand accumulation still continued on the Motosuka coast east of Katagai fishing port.

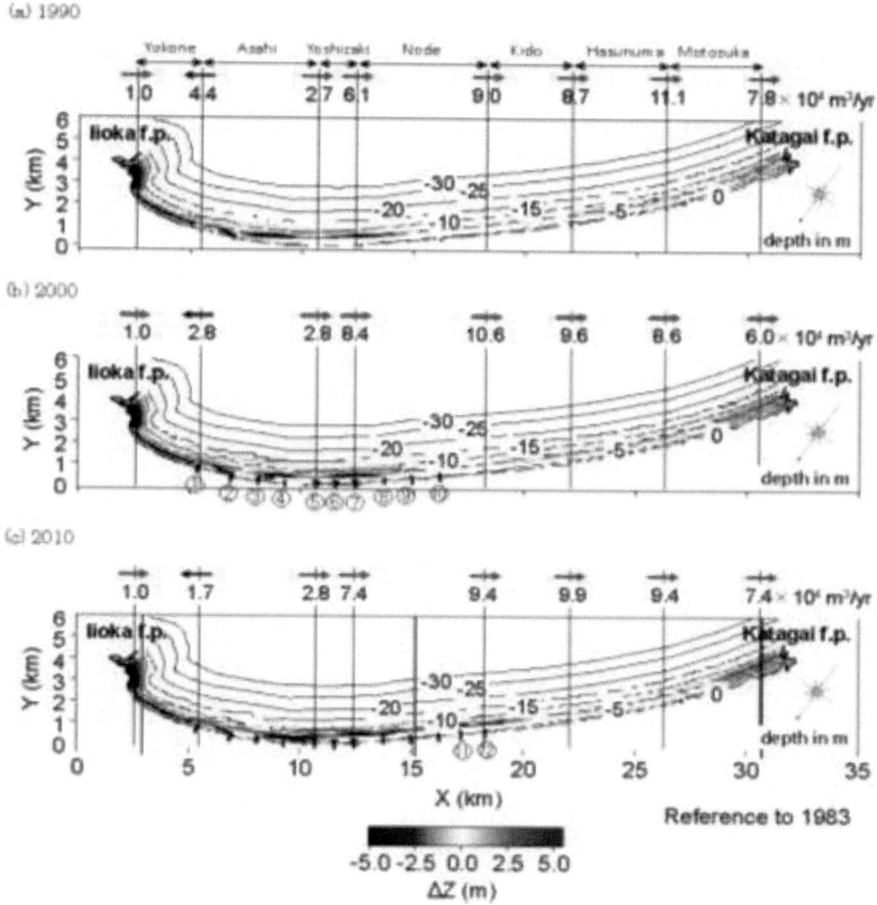

Figure 22. Reproduced bathymetries in 1990, 2000 and 2010.

On the Motosuka coast, the sand deposition zone expanded upcoast between 1990 and 2010, resulting in the monotonic increase in sand volume. The beach was eroded on the Hasunuma coast until 1990, but was accreted after 2000. This is because the Katagai fishing port breakwater was extended until 2000, and southwestward longshore sand transport was significantly obstructed by the breakwater. For example, southwestward longshore sand transport at a location of X = 31 km decreased from 7.8×10^4 m^3/yr between 1983 and 1990 to 6.0×10^4 m^3/yr between 1990 and 2000 owing to the extension of the Katagai fishing port breakwater. However, in 2010, longshore sand transport at a location of X =31 km increased again to 7.4×10^4 m^3/yr with an increase of 1.4×10^4 m^3/yr relative to that in 2000. This is considered to be due to the fact that sand was fully deposited in the area east of the Katagai fishing port breakwater and thus the ratio of obstructing southwestward longshore sand transport decreased.

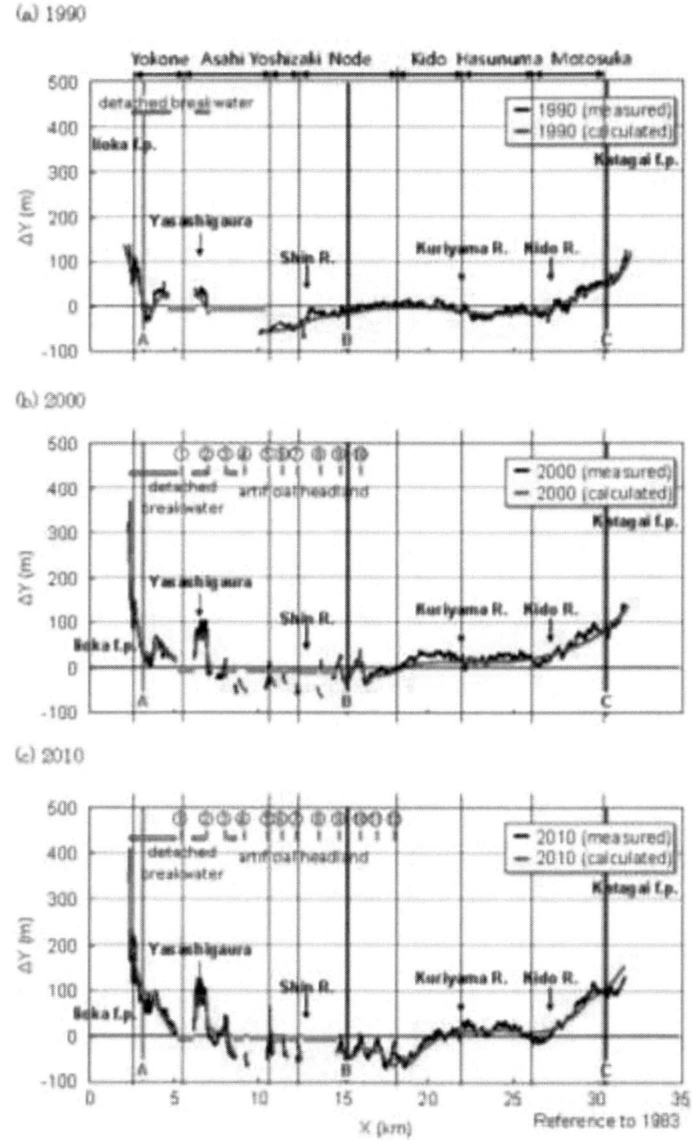

Figure 23. Measured and calculated shoreline configurations in 1990, 2000 and 2010.

Figure 23 shows the measured and predicted shoreline changes until 1990, 2000 and 2010 with reference to the shoreline in 1983. The shoreline advances measured on the Yokone coast west of Iioka fishing port and on the Motosuka coast north of Katagai fishing port are in good agreement with the predicted shoreline changes along with the shoreline advance behind the detached breakwaters at the Yashigaura coast. The predicted shoreline recession in the area west of the Node coast is also in good agreement with the measured changes.

DISCUSSION

Northern Kujukuri Beach has developed over several thousand years under the condition that a large amount of sand is continuously supplied from the sea cliffs of Byobugaura. This premise was completely lost owing to the construction of the breakwater along the sea cliffs to protect the tableland behind the sea cliffs since the 1970s. In addition, the Iioka fishing port breakwater had been extended at the north end of Kujukuri Beach since 1974, and longshore sand transport, which should be supplied to Kujukuri Beach was obstructed by the breakwater. As a result, the northern beach became eroded.

The south breakwater of Iioka fishing port had been extended up to a point of 1 km offshore of the original coastline by 1987, and therefore, a large wave-shelter zone was formed on the west side of the fishing port against the predominant waves incident from the southeast, inducing longshore sand transport from the outside to the inside of the wave-shelter zone and the resultant sand deposition on the Iioka coast.

In contrast to the sand deposition on the Iioka coast, the beach was eroded on the Asahi, Yoshizaki and Node coasts located south of the Iioka coast owing to the marked decrease in longshore sand supply from the upcoast, and the foreshore completely disappeared following the shoreline recession, exposing the seawall to waves. As a result of protecting the shoreline recession by using the seawall, beach erosion is severe further downcoast and the natural sandy beach is rapidly disappearing.

Although various measures including the construction of artificial headlands had been carried out, it became difficult to control beach erosion. Because Kujukuri Beach is composed of fine sand with a grain size ranging between 0.1 and 0.2 mm, and the equilibrium slope of the fine sand is as gentle as 1/40, even though the artificial headland is extended, longshore sand transport easily turns around the tip of the headland and is transported downcoast. Thus, the construction of artificial headlands is ineffective for controlling beach erosion.

On northern Kujukuri Beach, the seawall was exposed to waves along the coastline of 9.1 km length until 2010, i.e.,, one-third of the entire length of the sandy beach. The area with the exposed seawall is still expanding southwestward. Furthermore, the gently sloping revetments already constructed were often damaged by subsidence, and at each time, recovery works to the original form were carried out to maintain the present coastline. Thus, Kujukuri Beach with its natural scenic beauty has been lost.

Although the reconstruction of the damaged gently sloping revetment to recover the original shape has been carried out, the fundamental cause of the beach erosion cannot be eliminated by such works, and a chain reaction between the damage of the revetment and the recovery recurrently occurs with the downcoast expansion of the erosion zone. To solve the erosion problem fundamentally, increasing the sand supply is necessary. However, since the land behind the sea cliffs is also used extensively, it is impossible for the recession of Byobugaura to be permitted again. Also, it is difficult to cost-efficiently procure a large amount of sand from other areas. Because the sand volume deposited upcoast of Iioka fishing port is not large, sand bypassing using deposited sand is not effective, and sandback pass using the deposited sand at Katagai fishing port has a limited effect on the recovery of the sandy beach because of the limited volume of sand. Thus, the natural environment of Kujukuri Beach has been rapidly lost and an artificial coast covered with the seawall and concrete blocks has been formed.

The most fundamental cause of the beach erosion of northern Kujukuri Beach is the disruption of the continuous sand transport, which had continued for several thousand years, at the upcoast end of the littoral cell. When the continuous movement of longshore sand transport is disrupted, resulting in the downcoast erosion, measures using hard structures such as the seawall or detached breakwaters at the most severely eroded sites had been widely adopted in Japan. This allowed further downcoast erosion and the extension of the coastline protected by hard structures. In the example of northern Kujukuri Beach, the erosion zone has expanded downcoast at a rate of 270 m/yr. The setting back of the coastline as a measure against beach erosion is possible in principle. However, there is the coastal forest zone being subject to the Forest Law landward of the coastal protection zone under the jurisdiction of the Coast Law, and the management of activities by two independent authorities is normally difficult and time-consuming. In fact, such a measure is difficult to adopt.

CONCLUSION

In the example of northern Kujukuri Beach, shore protection was carried out only in a narrow band along the coastline using hard structures. Moreover, fishing ports were also constructed independently of the shore protection works without the comprehensive management of continuous sand transport in a wide area of a littoral cell. When many people realized the gravity of the erosion problem, it was already too late to fundamentally solve the issue. Thus, we cannot foresee a successful solution to the beach erosion of northern Kujukuri Beach at present. The only acceptable way is to concentrate on preventing the expansion of the erosion zone further downcoast, even though the most severely eroded sites covered with concrete blocks are left as they are.

REFERENCES

Horikawa, K. and Sunamura, T. 1969. Erosion of Byobugaura Sea Cliffs in Chiba Prefecture- investigation of recession of sea cliffs using aerial photographs (2), *Annu. J. Coastal Eng., JSCE*, Vol. 16, pp. 137-145. (in Japanese)

Horikawa, K. and Sunamura, T. 1970. Erosion of Byobugaura Sea Cliffs in Chiba Prefecture- investigation of recession of sea cliffs using aerial photographs (3), *Annual J. Coastal Eng., JSCE*, Vol. 17, pp. 289-296. (in Japanese)

Horikawa, K. and Sunamura, T. 1971. Study on direction of predominant sand transport on Kujukuri Beach in Chiba Prefecture, *Annual J. Coastal Eng., JSCE*, Vol. 18, pp. 417-422. (in Japanese)

Hoshigami, Y., Kobayashi, A., Uda, T. and Kumada, T. 2005. Recent formation and deformation of Kujukuri Beach, *J. Coastal Zone Studies*, Vol. 17, No. 3, pp. 47-56. (in Japanese)

Sakai, K., Uda, T., Serizawa, M., Kumada, T. and Kanda, Y. 2006. Model for predicting three-dimensional sea bottom topography of statically stable beach, *Proc. 30th* ICCE, pp. 3184-3196.

San-nami, T., Uda, T., Furuike, K. and Hoshigami, Y. 2009. Beach erosion of south Kujukuri Beach triggered by compound anthropogenic factors, *Asian and Pacific Coasts 2009, Proc. 5th Int. Conf.,* Vol. 4, pp. 252-258.

Serizawa, M., T. Uda, T. San-nami, K. Furuike and T. Kumada. 2003. Improvement of contour line change model in terms of stabilization mechanism of longitudinal profile, *Coastal Sediments* '03, pp. 1-15.

Serizawa, M., Uda, T., San-nami, T., Furuike, K., Ishikawa, T. and Kumada, T. 2007. Model for predicting beach changes on coast with sand of mixed grain size based on Bagnold's concept, *Coastal Sediments* '07, pp. 314-326.

Uda, T. 1997. *Beach erosion in Japan*, Sankaido Press, Tokyo, 442 p. (in Japanese)

Uda, T., Serizawa, M., Seino, S., Hoshigami, Y., San-nami, T., and Furuike, K. 2005. Summary of gently sloping revetment in Japan, *Proc. Int. Conf. Coastlines, Structures and Breakwaters, ICE*, pp. 1-10.

Uda, T., Sakai, K., Seino, S. and Hoshigami, Y. 2010. Formation of sandy beach and environmental change on upcoast of breakwater associated with extension of south breakwater of Iioka fishing port, *Proc. Civil Eng. in the Ocean, JSCE,* Vol. 26, pp. 1071-1076. (in Japanese)

Uda, T. 2010. *Japan's Beach Erosion - Reality and Future Measures*, World Scientific, Singapore, p. 418.

INDEX

D

E

F

G

H

I

J

K

L

M

N

O

P

Q

R

S

T